Computer Aided Design

The book comprehensively discusses principles, techniques, research activities, applications and case studies of computer-aided design in a single volume. The textbook will serve as ideal study material for undergraduate, and graduate students in a multitude of engineering disciplines.

The book

- Discusses techniques for wireframe, surface and solid modelling including practical cases and limitations.
- Each chapter contains solved examples and unsolved exercises.
- Includes research case studies and practical examples in enabling the user to link academic theory to engineering practice.
- Highlights the ability to convert graphic to non-graphic information such as in drawing up bills of materials in practice.
- Discusses important topics including constructive solid geometry, Boolean operations on solid primitives and Boolean algebra.

This text covers different aspects of computer-aided design, from the basic two-dimensional constructions through modifications, use of layers and dimensioning to advanced aspects such as three-dimensional modelling and customization of the package to suit different applications and disciplines. It further discusses important concepts including orthographic projections, isometric projections, 3D wireframe modelling, 3D surface modelling, solids of extrusion and solids of revolution. It will serve as ideal study material for undergraduate, and graduate students in the fields of mechanical engineering, industrial engineering, electrical and electronic engineering, civil and construction engineering, aerospace engineering and manufacturing engineering.

Computer Aided Design
Engineering Design and Modelling Using AutoCAD

Wilson R. Nyemba

CRC Press is an imprint of the
Taylor & Francis Group, an **informa** business

First edition published 2023
by CRC Press
6000 Broken Sound Parkway NW, Suite 300, Boca Raton, FL 33487-2742

and by CRC Press
4 Park Square, Milton Park, Abingdon, Oxon, OX14 4RN

CRC Press is an imprint of Taylor & Francis Group, LLC

Library of Congress Cataloging-in-Publication Data
Names: Nyemba, Wilson R., author.
Title: Computer aided design : engineering design and modeling using AutoCAD / Wilson R. Nyemba.
Description: First edition. | Boca Raton : CRC Press, [2023] | Includes bibliographical references and index.
Identifiers: LCCN 2022034547 (print) | LCCN 2022034548 (ebook) | ISBN 9781032265131 (hbk) | ISBN 9781032265148 (pbk) | ISBN 9781003288626 (ebk) | ISBN 9781032418858 (ebook plus)
Subjects: LCSH: Computer-aided design. | AutoCAD.
Classification: LCC T386.A97 N94 2023 (print) | LCC T386.A97 (ebook) | DDC 620/.00420285--dc23/eng/20221006
LC record available at https://lccn.loc.gov/2022034547
LC ebook record available at https://lccn.loc.gov/2022034548

ISBN: 978-1-032-26513-1 (hbk)
ISBN: 978-1-032-26514-8 (pbk)
ISBN: 978-1-003-28862-6 (ebk)
ISBN: 978-1-032-41885-8 (ebook plus)

DOI: 10.1201/9781003288626

Typeset in Sabon
by SPi Technologies India Pvt Ltd (Straive)

I sincerely dedicate this book to my family and all the engineering students and practising engineers to whom I have taught Computer Aided Engineering Design courses over the last 3 decades.

Wilson R. Nyemba

Contents

Preface

Computer Aided Design: Engineering Design and Modelling Using AutoCAD is a combined textbook and research monograph that was developed from various Computer Aided Engineering Design (CAED) courses and case studies carried out over the last 30 years. Academically, it is usually assumed that using a computer for drafting is generally faster than using a drawing board. However, this is not always true especially if the skills required to operate the computer and software are lacking, hence the main objective of the book is to equip users with appropriate skills that will enable them to use Computer Aided Design (CAD) systems effectively.

There is also a general trend in CAD that software packages such as AutoCAD are only used to produce 2D drawings. However, the concept of extending this into engineering design, 3D modelling and analysis is usually not fully exploited, thus companies do not fully realize the benefits of investing in CAD. While there are several CAD packages that can be acquired off-the-shelf, AutoCAD is one of those that can readily be customized and thus applicable for a multitude of disciplines, an aspect also covered in this book.

Although the book is based on AutoCAD 2021, it is presented in a flexible manner for users of other versions of AutoCAD. The various chapters cover different aspects of CAD from the basic two-dimensional constructions through modifications, use of layers and dimensioning to advanced aspects such as three-dimensional modelling and customization of the package to suit different applications and disciplines. As much as possible, the exercises in the book include the major disciplines in Engineering such as Mechanical, Civil and Electrical Engineering.

Various CAD textbooks available on the market focus on either engineering concepts or principles of engineering design in general, while the others assist users in operating different CAD systems such as AutoCAD, Solidworks etc. and are very specific to whichever CAD system would have been chosen. The novel approach in this book is that it contains a blend or cocktail of four aspects: principles of CAD and modelling, research and applications of CAD in academia and industry, its use as a manual or tutorial for AutoCAD 2021 and a guide for flexibility in transitioning from one version of AutoCAD to another or transitioning from one CAD system to another. The book also provides general guidelines to get the best out of investing in CAD, coupled with practical case studies but is by no means exhaustive of all that is available in AutoCAD, the rest of which is left for the user to explore further.

The book is primarily intended for undergraduate and postgraduate students pursuing various CAED courses. It is equally useful for design and practising engineers, drafting technicians and other personnel wishing to make the engineering design process an enjoyable and efficient one. After going through the various chapters, it is expected that users will have acquired sufficient skills to make good of their investment in CAD in order to boost their design and engineering operations. The book is also a precursor to the upcoming and more advanced one, **Computer-Integrated Engineering Design and Manufacture: Advanced Modelling and Analysis for Enhanced Engineering Operations.**

The e-book+ version of the book, Computer Aided Design: Engineering Design and Modelling using AutoCAD, complements the other versions of the book. The e-book+ version provides the user with additional features to enhance the user's understanding of this practical-based textbook. The four short videos recorded by the author and embedded in appropriate sections of the book outline and explain the key features of the book which include: an overall summary of the book, advanced aspects covered in the book such as 3D modelling and customization, the use of the book as a transitioning guide from previous versions of AutoCAD to newer ones or from one CAD system to another as well as practical applications in the form of case studies carried out by the author over the years.

Wilson R. Nyemba

Acknowledgements

I wish to acknowledge the support and interaction I had with undergraduate and postgraduate students over the last three decades. Initially, the AutoCAD software versions were those of the Disk Operating System (DOS) and gradually upgraded to newer and Windows versions over the years. These transitions also assisted in continually updating the book in its manuscript and draft form, till now, where it has been modified to provide flexibility in terms of the version of software being used.

I continued to interact with most of the graduate engineers but this time as practising design engineers in industry, with a quest for new knowledge, hence my continuous upgrading of the book and software. Their invitations to conduct refresher courses in-house, helped a lot in shaping the thrust of the book.

I am also indebted to my family for the sacrifice and time they allowed me to compile this book and above all the Almighty for giving me the strength and wisdom to eventually put it to print and publication after I released the first manuscript in 1996.

Wilson R. Nyemba

Author

Dr. Wilson R. Nyemba is a Senior Lecturer in Industrial and Mechatronics Engineering at the University of Zimbabwe with over 30 years of experience in both industry and academia. He designed and developed a wide range of engineering products in different capacities in the industry from product development to engineering management. He also formulated and led a number of successful ventures at the University of Zimbabwe where he served as Dean of the Faculty of Engineering. He also served as Chairman of WaterNet and Project Manager for the Royal Academy of Engineering projects for enhancing the quality of engineering education in Southern Africa. He is an accomplished consultant and researcher with interests in Computer Aided Engineering Design and Manufacture, Engineering Education, Capacity Building and Sustainability, Systems Engineering, Systems Thinking, Modelling and Simulation. He has taught Computer Aided Engineering Design and Manufacture, inclusive of CAD and CAM for over 25 years to undergraduate and postgraduate students in Engineering in general and Mechanical Engineering respectively, from where the material for this book was derived. He developed the original manuscript for this book in the mid-1990s specifically for Continuous Professional Development (CPD) for practising engineers in industry and has improved it over the years as the software also changed from one version to another, until this final manuscript of the book based on AutoCAD 2021 as the base software.

He holds a BSc Honours degree in Mechanical Engineering from the University of Zimbabwe, an MSc degree in Advanced Mechanical Engineering from the University of Warwick in England, funded by the British Council Overseas Development Authority Scholarship and a Doctor of Engineering degree in Mechanical Engineering from the University of

Johannesburg in South Africa, funded by the National Research Foundation (NRF) Doctoral Innovation Scholarship. He is also currently co-appointed as a Senior Research Associate in the Department of Quality Assurance and Operations Management at the University of Johannesburg. His consultancies, research and engagements with industry in Southern Africa over a period of over 10 years, culminated in the authoring of the book, *Bridging the Academia Industry Divide: Innovation and Industrialization Perspective using Systems Thinking Research in Sub-Saharan Africa* (ISBN 978-3-030-70492-6), published by Springer in 2021 and co-authored by Professor Charles Mbohwa of the University of Zimbabwe and Dr. Keith F. Carter of the University of Leicester in the UK. He has also carried out several consultancies in the area and published over 50 peer-reviewed papers in journals and conference proceedings. He has also received several awards and recognitions for outstanding research in his specific areas of research and interest.

Abbreviations

ADS	AutoCAD Development System
AME	Advanced Modelling Extensions
ASCII	American Standard Code for Information Interchange
ASD	Agile Software Development
ASE	AutoCAD Structured Query Language Extension
ASI	AutoCAD SQL Interface
BMP	Bitmap
BOM	Bills of Materials
B-Rep	Boundary Representation
CAD	Computer Aided Design
CADD	Computer Aided Drafting and Design
CAE	Computer Aided Engineering
CAGD	Computer Aided Geometric Design
CAID	Customized Add-on In-House Development
CAM	Computer Aided Manufacture
CDIO	Conceptualize, Design, Implement and Operate
CG	Computer Graphics
CIM	Computer Integrated Manufacture
CNC	Computer Numerical Control
COTS	Commercial Off-The-Shelf
CPU	Central Processing Unit
CSG	Constructive Solid Geometry
DCL	Dialogue Control Language
DLINE	Double Line
DOS	Disk Operating System
DPR	Dynamic Physical Rendering
DSDM	Dynamic Systems Development Model
DSETTINGS	Dynamic Settings
DWF	Design Web Format
DWG	Drawing Extension
DWT	Drawing Template Extension
DXF	Data Exchange Format
EDA	Electronic Design Automation

EPS	Encapsulated Post Script
GD&T	Geometrical Dimensioning and Tolerancing
GIGO	Garbage In Garbage Out
GPU	Graphics Processing Unit
GRAPHSCR	Graphics Screen
IBM	International Business Machines
IEEE	Institute for Electrical and Electronic Engineers
IEEM	Industrial Engineering and Engineering Management
IGES	Initial Graphics Exchange Specification
ISO	International Standards Organization
KPI	Key Performance Indicator
LISP	List Processing
LMB	Left Mouse Button
MDA	Mechanical Design Automation
MIT	Massachusetts Institute of Technology
MLINE	Multi-Line
MLSTYLE	Multi-Line Style
NC	Numerical Control
OBJ	Object (extension for 3D printable file)
ODF	Open Document Format
PDF	Portable Document Format
PDO	Protection Drawing Office
PGP	Program Parameters
RMB	Right Mouse Button
SNAPANG	Snap Angle
STL	Standard Triangle Language
TEXTSCR	Text Screen
UCS	User Coordinate System
WMF	Windows Meta File
XP	Scrum and Extreme Programming

Chapter 1

Introduction

1.1 BACKGROUND AND HISTORY OF CAD

The production of engineering drawings or any such representations dates back to the first industrial revolution. As the transformations in industry intensified, particularly between the second and third industrial revolutions (Nyemba et al., 2021), drawings and models were equally becoming complex, thus requiring quicker and better ways to generate, let alone manage them. This led to the development of computer-aided drafting and design (CADD) and computer-aided design (CAD) tools to aid in the creation or modification of drawings and analysis or optimization of designs (Narayan et al., 2008). The origins of CAD can be traced back to the mid-twentieth century when the first numerically controlled (NC) programming suite, PRONTO was developed by Dr. Patrick J. Hanratty, followed by SKETCHPAD developed by Ivan Sutherland at the Massachusetts Institute of Technology (MIT) and specifically meant for computer technical drawing (Dano, 2012). The early CAD systems were just meant to replace the monotonous drawing boards used in manual technical drawings. As such, this did not quite address the need for improved productivity in developing drawings, especially in view of the increased complexities. The other challenge was of course for the draftsmen and engineers to acquire appropriate skills in order to utilize the CAD systems.

One thing was certain though and that was the ease with which revisions and modifications on drawings could be achieved. More software developers came on board resulting in more affordable software packages and hardware in general, in terms of user-friendliness and functionality. To complement the 2D drafting capabilities of the available CAD systems, 3D wireframe features were incorporated a few years later with one of the earliest being Syntha Vision in 1969, becoming the first commercial solid modelling program. 3D Modelling was meant to enhance the visibility of designs and that in itself further enhanced the 3D capabilities, extending further to simulations, animations and finite element analysis. The software used in this book, AutoCAD was developed around the same time as the first release in 1982

DOI: 10.1201/9781003288626-1

but this was largely a disk-operating system (DOS) and of course quite cumbersome at that time. Further enhancements were introduced including parametric modelling in which models were defined by parameters, one of the most significant reasons for migrating from the drawing board to the computer, an aspect covered in Chapter 7.

In addition, and in line with parametric modelling, primitives were introduced to enhance the ease with which solid models were generated. These included boxes, cylinders, cones, pyramids etc. which were combined using Boolean algebra and constructive solid geometry (CSG) to produce any form of solid, as detailed in later Chapters 8–10. Autodesk developed AutoCAD using AutoLISP, a derivative of the LISP programming language. It allows users to write macro-programs that are well suited for graphics applications and thus enables customization of software to their specific needs. Unlike the latter part of the 20th century, the 21st century has witnessed the wide acceptance and application of CAD systems, equally with a wide variety of software, thus cheaper and more affordable. While the earlier versions of CAD were exclusive to 2D drafting, the situation has somewhat changed to where more time is spent on 3D modelling than 2D drafting (Narayan et al., 2008).

In addition, CAD systems are increasingly being integrated with computer-aided manufacturing (CAM) systems in such a way that whatever models are created in CAD, they can easily be transferrable to a CAM system for process planning and generation of NC files in preparation for manufacturing. Some experts predict that a lot is in store for the future in CAD, with anticipations such as '3D modelling will be closer to sculpting than painting' (Gherardini et al., 2018). A few years ago, no one ever thought it was possible to carry out 3D printing or additive manufacturing in the construction of 3D objects from CAD (Cummins, 2010).

In fact, in the 1980s, 3D printing was only suitable for the production of aesthetic prototypes, hence the term rapid prototyping at that time. This has been accomplished with ease and is now widespread in the production of various objects where the material is deposited, joined or solidified under a computer-controlled environment. The products generated from 3D printing are now widespread, including prosthetic limbs and body parts, prosthetics for Special Olympics runners, buildings, food, firearms and military equipment, musical instruments, to virtually anything that anyone can imagine (Cummins, 2010). More recently, the precision, repeatability and material range of 3D printing have been broadened to the extent of taking additive manufacturing as an industrial production technology (Lam et al., 2019). The rate at which transformations in 3D design and modelling have progressed only serves to prove that the future is ever so brighter. However, while some of these future predictions are optimistic, the future in technology can be very difficult to visualize except to postulate. Everything will depend on demand and the course of action that technologists may decide to take. However, what is certain about the future of CAD is that companies will adopt the standardized CAD based on the open document format (ODF) for ease of data exchange and the avoidance of duplication of tasks.

As the demand for new products increases and so does the number of designers and manufacturers, CAD software and hardware have been developed to increase the productivity of designers while improving the quality of designs and communication through electronic documentation (Narayan et al., 2008). Electronic CAD files are not useful for preparing documentation for intellectual property rights and inventions used in patenting but can also be easily converted for use in other software packages. Such use in designing electronic systems is commonly referred to as electronic design automation (EDA) while in mechanical design it is referred to as mechanical design automation (MDA) (Madsen and Madsen, 2012). MDA generally employs either vector-based graphics to represent the traditional drawing primitives such as lines, arcs etc. in manual drawing. However, the output in CAD goes beyond just showing the shapes and sizes but also includes vital information such as materials, processes, dimensions and tolerances according to the specific conventions in the particular application.

CAD has wide and extensive applications in industry for automotive design, shipbuilding, aerospace, architectural designs, prosthetics, parametrics, mine designs, etc. Other than the static models and designs, CAD has also been employed to produce simulations of the objects in real life as well as in the analysis of strengths and prediction of the performance of objects using finite element analysis. CAD has therefore become a major driver and tool for research in computational, graphic and differential geometry (Pottmann et al., 2007). Quite often such use of computers in geometrics is often referred to as computer-aided geometric design (CAGD) (Farin, 2002). Most young learners of CAD are introduced to the fundamental principles with a lot of enthusiasm to be able to get over the cumbersome use of the drawing boards and T-squares. Equally, practising engineers also push hard for their companies to invest in CAD and at times even in software and hardware that they may have been advised to be more efficient. For both groups of CAD users, the truth may not be far from that.

While so many reasons have been advanced for the need to invest in CAD, the most significant and immediate advantage or reason why technical personnel migrate from manual drawing boards to CAD is the ease and speed with which editing or modifications of drawings or designs can be carried out. However, these various reasons and advantages may not be achievable unless the appropriate skills to operate the CAD systems are available, hence the main objective of this book. A gradual and focused concentration of the guidelines provided in this book will help to amass the skills and techniques required to realize the full benefits of investing in CAD. It must also be noted that it may not be just a question of going through the various exercises and tutorials in this book but to master the techniques in order to nurture the skills. Just like learning how to drive is a technical skill that can be realized after several attempts, so is the use of CAD, and hence, repeated practice of the exercises is recommended. After acquiring the requisite skills to operate CAD systems, companies and institutions can realize a good return on their investment through various advantages such as the following:

1. Parametric designs: Most machines in the industry basically consist of standard components such as fasteners (bolts, nuts, screws, etc.). Through the use of blocks, these need only be created once and stored in libraries where they can be retrieved whenever necessary for other drawings or models.
2. Most CAD systems can be used to produce any kind of engineering or technical drawing that can be done by hand.
3. CAD systems are capable of producing technical drawings and models very quickly once one has grasped the techniques required. With experienced CAD draftspersons, speeds such as 10 times that are done manually are quite common.
4. Using CAD systems is less tedious and more accurate especially for such things as hatching, producing standard engineering components or automatic dimensioning.
5. Drawings or parts of drawings can be copied, scaled, rotated, mirrored, or moved with ease. They can also be inserted into sections of other drawings with ease, unlike in manual drafting where everything has to be repeatedly carried out unless templates are employed to ease the repetition.
6. New details can be added to a drawing, or detail within a drawing can be altered with ease without having to make any mechanical or manual erasers.
7. Drawings can be dimensioned automatically with accuracy, greatly reducing the possibility of dimensional error.
8. The production of complex drawings, with many details on limited paper sizes, can be cumbersome. However, the use of layers in CAD allows users to place selected details on different layers and in turn, the use of layer control tools such as turning layers on and off, freezing and thawing as well as locking and unlocking helps to clear the clutter as well as reduce the risk of accidentally deleting important details.
9. One of the major challenges in managing or handling manual drawings produced on paper is that of storage. Prior to the introduction of CAD, many companies had to invest in numerous storage cabinets that took up a lot of space. Nowadays, micro-SD cards, the size of a fingernail are capable of holding several gigabytes of data and thus can easily hold many drawing files, as compared in Figure 1.1, saving a lot of space for the company or institution.
10. Drawings need only be plotted or printed when it is absolutely necessary, otherwise engineers and artisans can even their electronic gadgets to view any drawings on site, saving the company or institution on paper and printing costs.

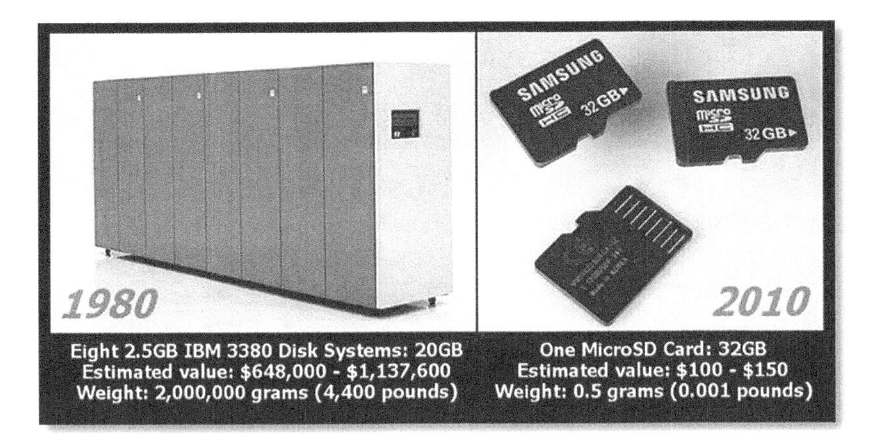

Figure 1.1 Comparison of storage between 1980 and 2010.

11. Drawings produced in CAD software can be transferred and used in other CAD software, applications or databases using such tools as DXF or IGES.

1.2 HARDWARE AND SOFTWARE REQUIREMENTS AND DEVICES

Generally, hardware for CAD systems can be broadly classified into three categories: Input devices, Processors and Output devices. From inception, these devices were supplied by only a few manufacturers but the transition from the Third to the Fourth Industrial Revolution has seen more manufacturers and developers providing a wide range of such devices as well as increased computing power and capabilities (Sharma & Singh, 2020). The sprouting of these suppliers and global competition also meant that the developers were not only forced to reduce prices in order to remain competitive, hence the availing of both hardware and software to more users, but the gadgets have also increasingly been made smaller and more user-friendly.

1.2.1 Input devices

Several devices can be used to input information into CAD systems. However, three basic requirements are shown in Figure 1.2. The mouse can be wireless or wired but should have a left mouse button (LMB), a right mouse button (RMB) and a roller in between. A standard keyboard would be sufficient, preferably one with function keys for certain functions in specific CAD software. The digitizer is optional but very useful in the conversion of manual to electronic drawings.

Figure 1.2 CAD systems input devices.

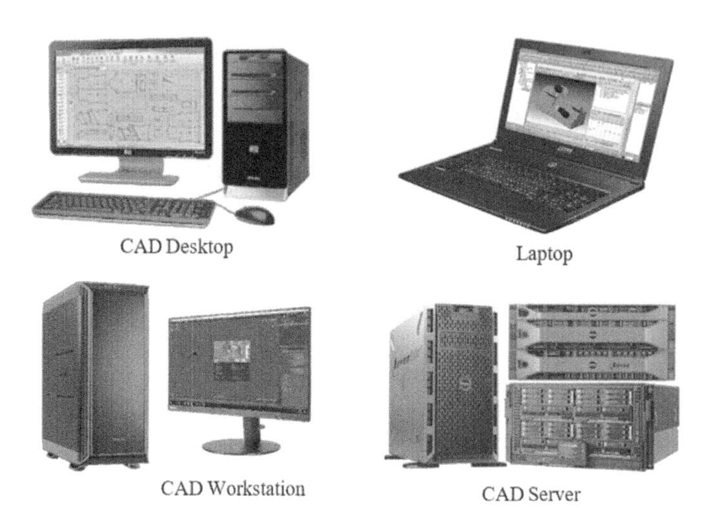

Figure 1.3 CAD processors.

1.2.2 Processors

CAD systems can be operated from a wide variety of processors supplied by different manufacturers and these can be laptops or desktop PCs mainly for the individual user or CAD workstations or servers for large corporates, as shown in Figure 1.3. The minimum specifications for such processors have been changing rapidly and in tandem with the dynamic changes in technology. However, as of 2021, Autodesk, the developers of the software used in this book, recommend the minimum specifications shown in Table 1.1 (Autodesk, 2021), under the Windows environment employed in this book.

1.2.3 Output devices

The most important output device for any CAD system is the display screen. In tandem with developments in technology, at least a 17-inch coupled with 1920 × 1080 display with true colour and high resolution as detailed in Table 1.1 is recommended. In some instances, some corporates combine such displays to obtain quad monitors for clarity, as shown in Figure 1.4. Only if it is absolutely necessary, printers and plotters as shown in Figure 1.4 can

Table 1.1 Windows system requirements for AutoCAD 2021 (Autodesk, 2021)

Operating system	Microsoft® Windows® 7 SP1 with Update KB4019990 (64-bit only) Microsoft Windows 8.1 with Update KB2919355 (64-bit only) Microsoft Windows 10 (64-bit only) (version 1803 or higher)
Processor	**Basic**: 2.5–2.9 GHz processor **Recommended**: 3+ GHz processor, Multiple processors: Supported by the application
Memory	**Basic**: 8 GB, **Recommended**: 16 GB
Display resolution	**Conventional Displays**: 1920 × 1080 with True Colour **High Resolution & 4K Displays**: Resolutions up to 3840 × 2160 supported on Windows 10, 64-bit systems (with capable display card)
Display card	**Basic**: 1 GB GPU with 29 GB/s Bandwidth and DirectX 11 compliant **Recommended**: 4 GB GPU with 106 GB/s Bandwidth and DirectX 11 compliant
Disk space	6.0 GB
Browser	Google Chrome™ (for AutoCAD web app)
Network	Deployment via Deployment Wizard. The license server and all workstations that will run applications dependent on network licensing must run TCP/IP protocol. Either Microsoft® or Novell TCP/IP protocol stacks are acceptable. Primary login on workstations may be Netware or Windows. Operating systems supported for the application, the license server will run on the Windows Server® 2016, Windows Server 2012, and Windows Server 2012 R2
Pointing device	MS-Mouse compliant
.NET framework	.NET Framework Version 4.7 or later *DirectX11 recommended by supported OS

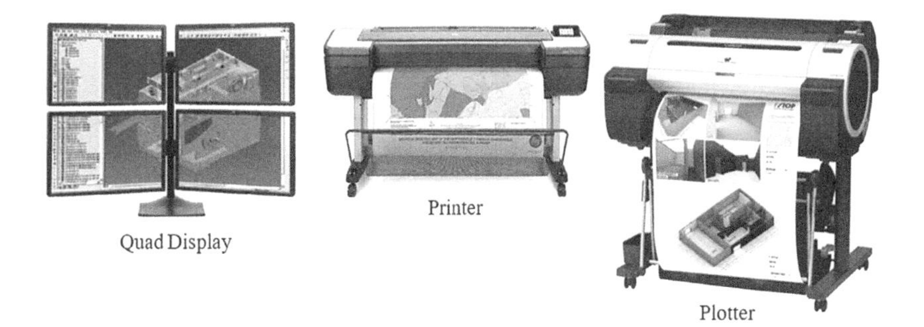

Figure 1.4 CAD output devices.

also be used, otherwise the trend nowadays is to use display screens. Not only does this save on paper but also on printer/plotter toner and cartridges.

1.2.4 Software

Unlike the mid-19th century when CAD was in its infancy and there were very few options in terms of CAD software and operating systems, many software developers have emerged in recent years. This has resulted in varying software packages which are now readily available off-the-shelf. Some of these are complete suites that allow for designs in various disciplines and some have been designed specifically for certain tasks and disciplines in engineering. Figure 1.5 shows some of the widely used CAD systems.

The usefulness of each package is dependent on what it is capable of doing, user-friendliness and ease of use, customization capabilities, electronic files interchangeability, cost, and customer support among other considerations. According to Carlota (2019), the top-10 CAD packages are as

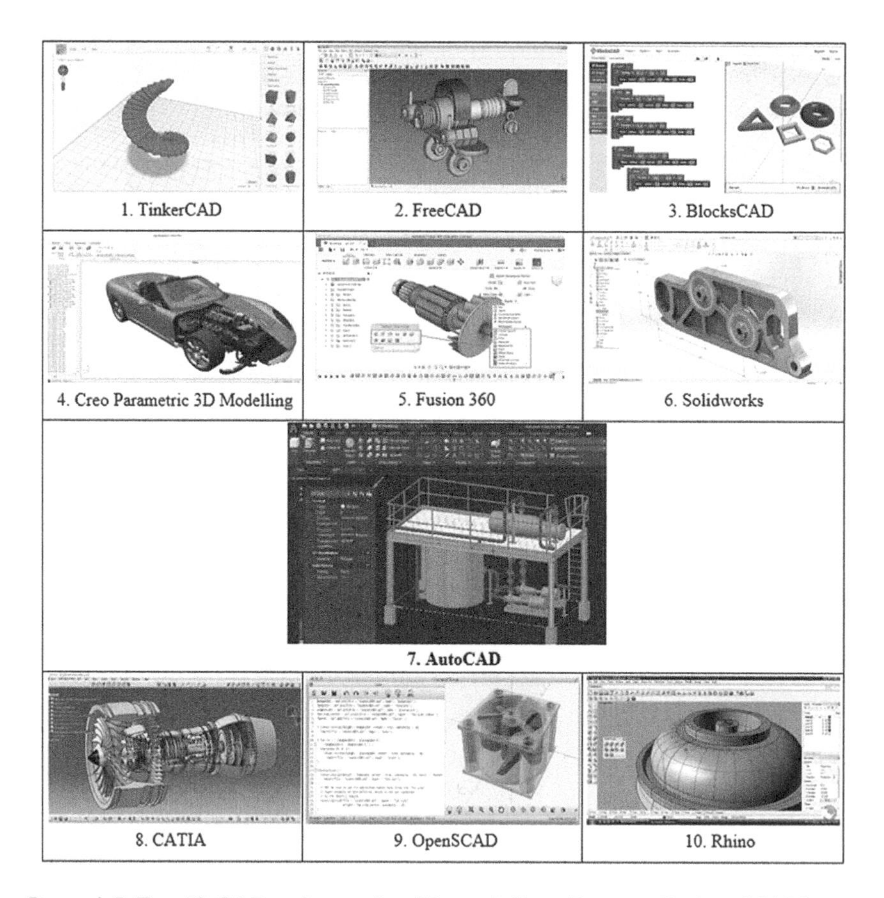

Figure 1.5 Top 10 CAD software for 3D modelling. (Source: Carlota 2019.)

shown in Figure 1.5. The magnification of AutoCAD is simply an indication of the software used in this book. Many of these commercial and proprietary programs have their pros and cons and also have strong open-source alternatives. Most of them were developed for specific purposes. The above top-10 list has changed over the last few years because of users' changing preferences. While previously, AutoCAD was in position 3 in 2018, it has moved down to position 7 in 2021. Previously, CAD systems such as TurboCAD Deluxe, IronCAD, Shapr3D and BricsCAD all featured in the top 10 but have since been overtaken by other newer packages.

Traditional packages such as TinkerCAD, AutoCAD Solidworks and Fusion 360 have all remained within the top bracket. Traditionally, the earlier CAD systems operated on DOS before the advent of Windows but more recently, all these packages now operated mostly under Windows or Linux. Although Solidworks and AutoCAD are down to positions 6 and 7, these are probably the most widely used CAD packages the world over (Carlota, 2019). The future in CAD systems as the available developers compete against each other, will be in those that can easily be integrated with CAM, particularly 3D printing and additive manufacturing. In addition, cost, user-friendliness, adaptability and capabilities for customization will be key parameters and considerations for future CAD systems. The future is also projected to witness more development of hardware and software for rapid manufacturing under CAM and rapid prototyping under additive manufacturing. This will revolutionize manufacturing, marketing and business processes in general. The development of more options for 3D printing, laser and metal sintering will boost the production of complex parts efficiently. This is anticipated to reduce lead times for products significantly.

CAD systems have been used to develop a wide spectrum of facilities to aid and make the process of product development to manufacture easier and more cost-effective. These include the capabilities for dynamic physical rendering (DPR) and the evolvement of new representations of 3D models, as in the little Lego blocks used to build different models, commonly referred to as claytronics (Follmer et al., 2013). Although the use of CAD in genetic programming may be considered weak at the moment, the increased use and development of machine learning uses evolutionary algorithms to optimize sets of designs. Genetic programming finds wide and successful applications in the development of computer programs, electronic digital circuits or transmitting devices such as aerials. It is most suitable for the design of complex objects or products which consist of many parts. Thus, CAD systems will incorporate genetic programming methods to improve and speed up the production of hydraulic and fluid control systems, for example. Effectively, computer programs generated using genetic algorithms will serve like DNA, while manufacturing hardware will serve as ribosome (complex molecular machine found in the cytoplasm of living cells where proteins are synthesized), thus allowing one machine to manufacture a wide range of different products (Wang et al., 2020).

1.3 BACKGROUND TO AUTOCAD

AutoCAD is a general purpose and commercial CADD package developed in 1982 and marketed by Autodesk. Before its introduction, many earlier CAD packages were run either on mainframe or mini-computers under the DOS with each operator working from a separate graphics terminal. Over the years, several versions of AutoCAD have been developed and released to the market periodically. In the 1980s, newer versions could be released after four or five years. However, in recent years, almost every year, Autodesk releases a new version of AutoCAD.

The new versions released each time are invariably an improvement over previous versions but the basic principles and operations remain the same. In addition, some of the newer versions come with additional add-on facilities to enhance the design of components or products for specific applications in the various disciplines of engineering (civil, mechanical, electrical, etc.) and the built environment (architecture, quantity surveying etc.)

Unlike many of the CAD packages available on the market such as those in Figure 1.5, AutoCAD is one of the many software packages developed by Autodesk, some of which are Assemble, 3Ds Max, 360 Pro, Fusion 360, Inventor, Plant 3D, Piping and Instrumentation Design, etc. As the names of these specific packages depict, the various Autodesk packages were developed to address specific tasks in the various sub-disciplines such as 3D modelling, instrumentation, pipe networks, etc. However, the general-purpose AutoCAD package comes with almost all the capabilities found in other Autodesk products. As such, it is one of the most flexible packages available, making it the most ideal for use in training engineering students in various disciplines. The other CAD packages listed in Figure 1.5 are task-specific and may not be as flexible as the general-purpose AutoCAD package. In addition, and as such, the following are some of the applications for which the general-purpose AutoCAD can be used, in particular, for the training of future personnel in these areas:

1. Architectural drawings of all kinds including interior designs and facilities
2. Work-flow charts and organizational diagrams
3. Electronic/electrical wiring, chemical and civil hydraulics
4. Mechanical, automotive and aerospace engineering applications
5. Mining and mineral processing plants and process flows
6. Topographic maps and nautical charts for surveying and geomatics
7. Plots and other representations of mathematical and scientific functions
8. Technical illustrations, machine designs and assembly diagrams
9. Limited simulation and analysis of linkages and mechanisms
10. 3D modelling and analysis
11. Parametric designing

The recent versions of AutoCAD have blended in very well and mostly operated under the Windows environment in an interactive, menu-driven way, and are designed to be reasonably easy to learn and use. Ready-access help and reference guides for commands and other operations are provided to help the user whenever necessary. The complete guide through this book, based on AutoCAD 2021 will cover all the basic fundamental and basic principles and will also introduce advanced aspects such as 3D modelling, which is covered in more detail in the other upcoming book on CAD, CAE and CAM. Examples and tutorials provided in this book will enable users, from engineering students to engineering practitioners to gain experience with the standard features of the software. The knowledge and experience acquired by users will be versatile enough for them to be able to operate other CAD systems such as Solidworks.

1.4 LIMITATIONS TO THE USE OF CAD

While there are many reasons why companies and institutions should seriously consider investing in CAD for their students and engineers alike, it should be borne in mind that, such investments must be done carefully and gradually. Just like any other computer software and tools, CAD systems require appropriate skills to operate effectively and efficiently, hence this is the main objective of developing this comprehensive book for CAD practitioners. CAD systems are not fully automated to the extent that they can run unattended such as some of the processing plants in manufacturing nowadays. The user needs to be fully equipped to be able to 'drive' the system appropriately, otherwise as the saying goes, 'Garbage in garbage out' (GIGO). In addition, it is also equally costly to train personnel as this is almost a must to be able to cope with the frequent changes in the CAD software packages. However, the costly part is usually the start and any changes in software can easily be adapted by well-trained personnel.

Although computer hardware and software costs have been going down due to global competition, some countries especially those in the industrializing world do not always have sufficient financial resources to either replenish their hardware or update their software. Generally, this presents a challenge for sometimes operating with hardware and software that may not closely match, that is, either old computers with new software or new computers with old software. This buttresses the need to make investment decisions of this nature carefully and perhaps take a gradual approach. The other challenge in using CAD systems is the mismatch of technologies between design and manufacture. Due to insufficient resources, companies or institutions in industrializing countries seldom have matching equipment for design and manufacture, such as having all the software for CAD and CAM but no computer numerically controlled (CNC) to upload the post-processed NC files for manufacture. Investment in modern technology can be quite capital intensive and hence careful planning is required.

1.5 BASICS FOR USING AUTOCAD

Unlike most of the earlier CAD systems that were largely run through DOS, recent versions are now operated under either the Windows or Linux environment. This book is based on AutoCAD 2021 operated under the Windows 10 Professional environment by clicking on the desktop icon as shown in Figure 1.6.

1.5.1 AutoCAD screen

Figure 1.7 shows the standard AutoCAD 2021 screen. AutoCAD uses dual display screens in the form of the Graphics Window (Drawing Editor) and

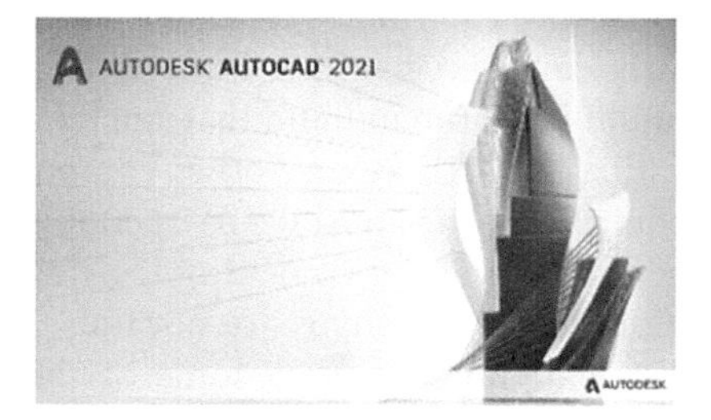

Figure 1.6 Starting AutoCAD 2021.

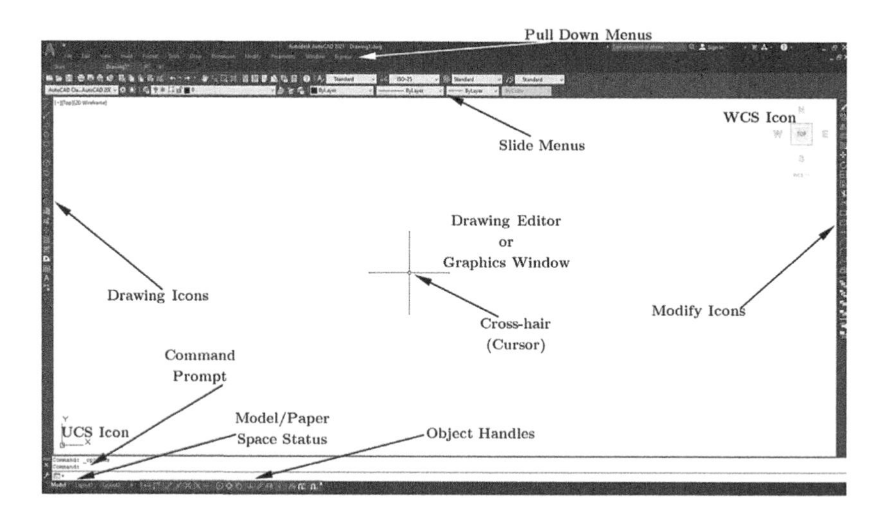

Figure 1.7 AutoCAD 2021 screen.

Command Prompts (Text Output). However, on some computers, a single PC serves for both graphics and text. By default, when AutoCAD starts, it displays the graphics window and only three lines of the text screen at the bottom (Command Prompt area). The rest of the text lines lie behind the graphics window. To access these lines or flip between the two screens, the commands TEXTSCR and GRAPHSCR can be used. The older versions of AutoCAD used the function key F1 for this purpose.

While Figure 1.7 shows the standard display with minimal icons and other menus, there are numerous other icons that can be displayed for ease of use. However, it is not advisable for beginners as the clutter on the screen may add to the confusion and thus defeat the whole purpose of improving productivity in the development of electronic drawings.

In addition, the use of icons requires one with experience to manoeuvre through the different icons available. As such, it is also advisable to get the grasp and understanding of various commands and their aliases provided by AutoCAD to such an extent that a complete drawing can be developed using the command prompt. The advantage of using the command prompt is that as the versions of AutoCAD are continually upgraded, the commands will remain the same but to a large extent, the display screen, icons and slide menus may change with changes in the versions. In some instances, certain menus and icons are removed completely. Thus, the knowledge of the AutoCAD commands and command aliases is vital for easy transitioning from one version to another.

To insert additional icons or menus, right-click on the blank area adjacent to the slide menus and select AutoCAD from where preferred icons to display can be selected. Normally, AutoCAD uses two display screens; one for command prompts and text output and the other for graphics. However, on many PCs, a single PC serves for both graphics and text. In the Drawing Editor, AutoCAD holds the last 24 text lines in reserve and displays only the last three text lines by pressing the F1 key, (FLIP SCREEN). In this way, AutoCAD behaves as if it had two display screens at its disposal.

1.5.2 Keyboard and mouse

The basic input devices for CAD systems including AutoCAD are the keyboard and the mouse. While the standard keyboard is adequate, the following are important keys that should be remembered when operating AutoCAD in order to speed up productivity in the development of drawings. The <enter> key acts as the communication link between the user and the system. This key can be used after typing a command for AutoCAD to process. The <backspace> key is used to point at menu items or to locate objects in the drawing area and to allow the user to continue executing a previous command.

The cross-hair or cursor can be moved from one position to another, not only by the mouse but by the arrow keys on the keyboard as well.

The <page up> key and <page down> keys can be used to speed up or slow down the cursor respectively. The mouse has three key features that are necessary for AutoCAD, that is, the LMB for the selection of points on the graphics window to pick an icon or menu item. Although not frequently used, the RMB is used to quickly access Edit commands as well as previously used commands anywhere on the graphics window. The roller in between the LMB and RMB can be used to scroll up or down or to zoom in and out. The touchpad on the laptop can also be used for the same tasks.

1.5.3 Object handles

There are various facilities that have been included in AutoCAD to enable the system to produce drawings with ease. These are generally referred to as object handles. These include *endpoint, midpoint, centre, intersection, quadrant, tangent, perpendicular*, etc. They are also assigned appropriate symbols as shown in Figure 1.8. These object snap icons can also be turned on by right-clicking on the empty space adjacent to the right of and adjacent to the slide menus and selecting AutoCAD and then *Object Snap*. In addition, the SNAP command can also be used to 'snap' all input coordinates to the nearest point on the snap grid by specifying the snap spacing between grid points. The GRID command can also be used to set the desired grid spacing. It is usual to set the grid to be either the snap spacing or twice the snap spacing, while the same command is used to turn on the grid to display dots on the graphics window to enable the user easily identify a point in relation to the previous one.

The ORTHO command is used to enable the cursor to draw only horizontal or vertical lines, parallel to the x-, y- or z-axes or lines that are parallel to the isometric planes as described in detail in Chapter 6. Alternatively, this can be turned on temporarily by holding down the <shift> while drawing.

COORD command allows users to specify the format of the coordinate system display, which forms part of the status line in the drawing editor. There are three choices: display of absolute coordinates is continuously updated as the cursor moves, the display of absolute coordinates is only updated if a point is picked and the distance and the angle from the last point are updated continuously when a rubber band cursor is on the screen.

Primitives such as lines, arcs etc. can be drawn using absolute (x, y, z), relative (Dx, Dy, Dz) or polar (reference point and angle of inclination to the next point) coordinates by invoking the DSETTINGS command to set

Figure 1.8 Object snap modes.

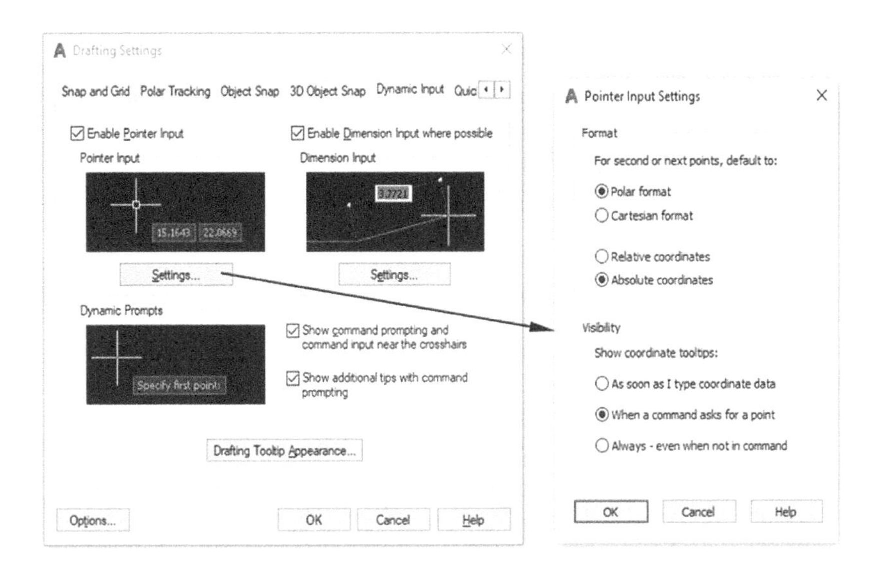

Figure 1.9 Dynamic settings for input.

the desired input method by choosing *Dynamic Input* from the dialogue box as shown in Figure 1.9. It is advisable to keep *Polar* and *Absolute* ON but not all of them to avoid confusion on dynamic input.

1.5.4 Methods of entering commands

There are four ways in which AutoCAD can be instructed to execute commands, that is, command prompt, pull-down menus, icon/slide menus or the mouse. However, as pointed out earlier, it will be very good practice to master and adopt the use of the command prompt as the list of commands does not vary with changes and updated versions of AutoCAD, as much as icons and pull-down menus may change.

Most AutoCAD commands have been shortened to command aliases for ease of use. These include LINE – L, CIRCLE – C, MOVE – M, COPY – CP, etc. and these can be found in the reference guide/manual or simply selecting the question mark (?) on the slide menu or pressing the function key F1 for help. The drawing of various building blocks or drawing primitives such as lines, arcs, circles etc. can also be accomplished by accessing icon menus located to the left of the graphics window as shown in Figure 1.7. The default icon menu is not exhaustive but contains the most commonly used primitives. Although it is located on the left side of the screen by default, users have the flexibility to drag this to any position they wish, for ease of use or ergonomics. In addition, the icon menus are frequently changed with new versions of AutoCAD, either by adding extra icons or removing those that may no longer be used or no longer part of AutoCAD such as what happened with the double line (DLINE) in the earlier versions of AutoCAD. The icon menus can also be customized to contain preferred icons or complete new icons can be customized for use at specific companies, an issue that is dealt with later in Chapter 11.

Most of the 12 pull-down menus from *File, Edit, View, Insert* .... *Express* contain standard Microsoft Windows pull-down menus that are used in many different software packages running under the Windows environment, with exception of those like *Draw, Dimension, Modify, Parametric and Express*, which are exclusively designed for use in AutoCAD.

Drawing primitives can also be produced by selecting the *Draw* pull-down menu to access *Line, Polyline, Arc, Circle*, etc. As with the icon menus, these also require familiarity to speed up the production of drawings, an attribute acquired with experience. Both the icon and pull-down menus can be accessed using the LMB of the mouse, which can equally be used as a fourth method for entering commands in the system especially when the RMB can be used to retrieve previously executed commands or the *Edit* menu.

1.5.5 Specifying coordinates/points to the drawing editor

Generally, there are four ways in which points or coordinates can be specified on a drawing editor, that is, Absolute Cartesian Coordinates, Relative Coordinates, Polar Coordinates and the mouse for rough sketches that do not need specific coordinates.

Absolute Cartesian Coordinates

If a particular point or coordinate is expressly known, then the (x, y, z) values can be entered when required. For 2D drafting, only the first two values are necessary as the z coordinate is assumed to be zero.

Relative Coordinates

Relative coordinates are derived from a reference of the existing point from where the variations in values of the new position are used to define the new point or position, that is, (Dz). Similar to absolute coordinates, specifying just two numbers separated by a comma, implies that the change in z (Dz) is zero. For example, if a given reference point is (20.29,36.14) and another point is $(Dx, Dy) = (3,2)$, then this can be specified by prefixing the changes in x and y by the @ symbol, that is,

@3,2 to obtain the coordinates of the new position as (23.29,38.14)

Polar Coordinates

Polar coordinates can also be used to specify a point on the graphics window, *relative* to a given reference point but this time, the distance between the reference point and the angle of inclination (taken anticlockwise) are

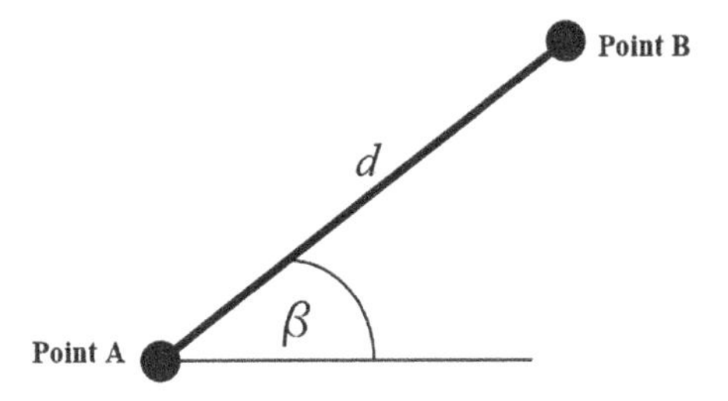

Figure 1.10 Polar coordinates.

required. For example, if it is given that the absolute coordinates for point A are known as shown in Figure 1.10:

Assuming that the point *B* is inclined at an angle b, taken positive, anticlockwise from the horizontal and that the distance between the two points is *d*, then point *B* relative to point *A* can be specified using polar coordinates by prefixing the response to the command by @ and also including the angle of inclination using the < symbol as follows:

Point B relative to Point A is given as:
@d<b

1.5.6 Setting drawing limits and scales

As CAD drawings are generated, it is strongly advisable to develop them using a full scale, the standard of which metric or millimetres (mm) are adopted in engineering work. The scaling of the drawings to fit whatever size of paper is desired is usually done as the last step when preparing such drawings for output (printer or plotter).

Scaling of the drawings at the model development stage can result in undesirable consequences for outputs. Manual drawings have traditionally been produced on different sizes of paper generally ranging from the small size (A4) to the largest (A0). The dimensions for the various paper sizes in landscape format are as proved in Table 1.2. Depending on the desired output or paper size, the drawing limits for the graphics window can be set using the

Table 1.2 Drawing paper sizes

Paper size	*Length/mm*	*Height/mm*
A4	297	210
A3	420	297
A2	594	420
A1	840	594
A0	1188	840

LIMITS command and then specify the lower left corner as (0,0) and the upper right corner as per dimensions in Table 1.1. Before developing any drawings after setting these limits, it is also advisable to ZOOM–EXTENTS to bring the entire paper space onto the screen which can be zoomed in or out as desired as the user focuses on certain portions of the drawing.

After zooming to extents, the 'paper' limits can be checked by hovering around the bottom left corner of the drawing editor to confirm the (0,0) coordinate and also the top right corner to confirm the coordinates according to the paper size set. The default limits set by AutoCAD are for the A3 paper (420,297). However, to see these coordinates on the screen is dependent on the size and aspect ratio of the screen and thus it may not be possible to see exactly (420,297) but close enough to confirm such.

It is also advisable to maintain full scale using mm as the base or standard of measurement, unless otherwise specified, for instance in some countries, the base or standard measurement can be imperial (inches). For close scrutiny and clarity, models or drawings can be zoomed in to the area under consideration but this should not be confused with scaling the drawing up or down.

Scaling magnifies the model or drawing by a chosen scale factor. For example, if a line is originally drawn as 100 mm and then scaled by 2, it becomes 200 mm long, an aspect which is more applicable at the printing stage. The visibility of certain lines such as dashed/hidden or dotted can be adjusted using the linetype scale command.

1.5.7 Types of lines used in AutoCAD

As with manual or conventional drawings produced using the drawing board such as *continuous*, *hidden* (*dashed*), *centre* (*dash-dot*), etc., AutoCAD comprises a comprehensive suite of such lines and these are accessible from the properties slide menu as shown in Figure 1.7. The default lines that are loaded when AutoCAD is started are the standard continuous lines, and additional ones can be loaded by selecting other as shown in Figure 1.11.

Other special lines that are not covered under this suite are polylines and multi-lines. The ordinary and standard lines in AutoCAD are of a fixed thickness or width and thus, their boldness cannot be adjusted. In order to produce lines of varying thickness or width, such as those used for the borders of figures or title blocks, the polyline is used. A polyline can be viewed as a stack of lines piled on top of each other, thus arriving at a desirable thickness. Using the PLINE command users can specify the width (thickness) from the beginning to the end of the line. Normally, this width is the same at the beginning as it will be at the end unless a tapered line is required.

AutoCAD also has the facility for drawing standard thickness lines that are parallel to each other. In the older versions of AutoCAD, this was referred to as the double line which was obtained using the DLINE command but in the later versions of the software, this has been changed to multi-line,

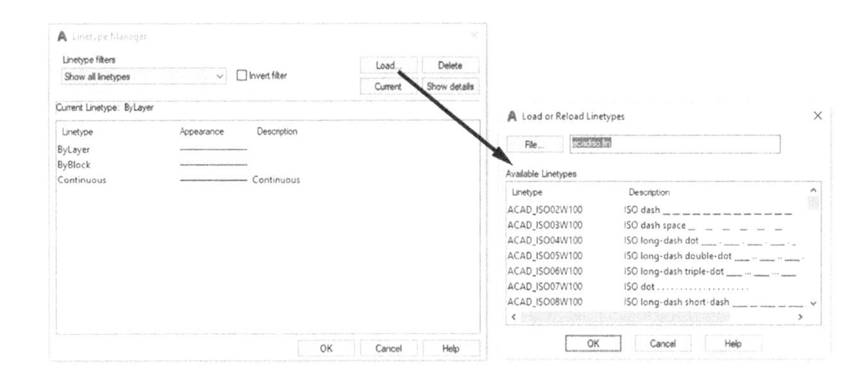

Figure 1.11 Loading the suite of lines in AutoCAD.

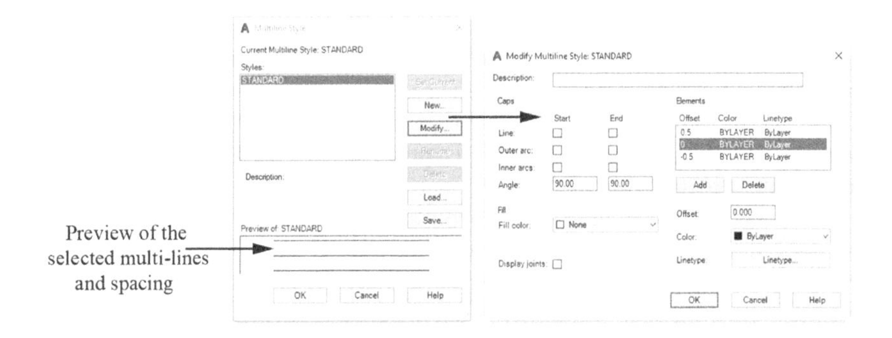

Figure 1.12 Multi-line styles dialogue box.

obtained by using the MLINE or ML command. The difference is that the user now has a choice of not only specifying the space between the lines but also how many lines can be drawn parallel to each other. All these settings, such as how many lines, spacing, capped at the beginning or end or both, can be set by using the MLSTYLE command to get the dialogue boxes as shown in Figure 1.12.

1.5.8 Saving drawings and file exchange formats

To avoid the loss of any models or drawings under development, it is advisable to periodically save them in a systematic way, such as defining the project(s) and the various subsections of the project(s) under appropriate folders and file names. However, most of the recent versions of AutoCAD offer the facility for auto-saving as the development of the drawing or model is in progress. In addition, the newer versions of AutoCAD also allow the automatic removal of unwanted objects, especially those created after trimming or erasing. In the earlier versions, the PURGE command was used

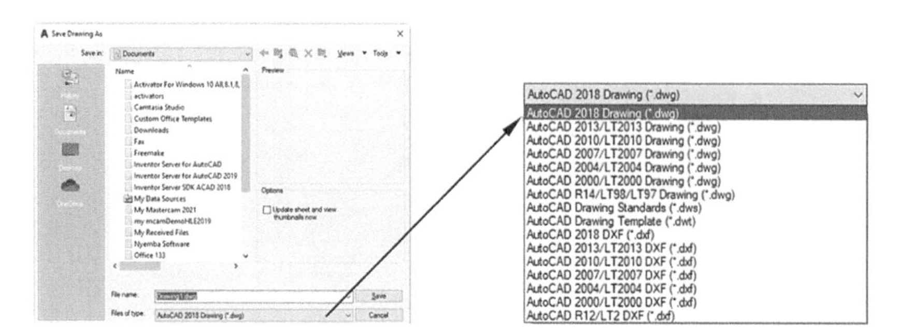

Figure 1.13 Various formats for saving AutoCAD drawings and models.

to erase such unwanted objects. Removal of such unwanted objects as the modelling is in progress helps to reduce or optimize the file size when saved.

The command QSAVE is equivalent to the *Save* command in other Windows packages where the model under development will be saved under the name in which it would have been created. On the other hand, the SAVE command in AutoCAD is equivalent to the *Save As* command in other Windows packages where the user is prompted to rename the file under development to some other name.

AutoCAD offers the users various options and formats to save drawings, the default being the ordinary drawing or DWG file such as Name.dwg. Other options include the DWT extension which allows users to save an ordinary drawing as a template for use when generating other drawings in future. Typically, this is the file extension used for generating title blocks for drawings which can later be retrieved and used to fill in any other new drawings. This will be dealt with in detail in a later chapter.

Files saved with an extension DXF are interchangeable data exchange format files for use in other Windows packages, including those saved as Windows Metafiles (WMF) or Bitmaps (BMP). Other formats and file exchanges include the Encapsulated PostScript file (EPS) and American Standard Code for Information Interchange (ASCII) which uses the SAT extension. With the advent of the internet and the need to post some files on websites, Autodesk has also introduced the Design Web Format (DWF). Another useful format is the Portable Document Format (PDF) which will be dealt with in more detail in Chapter 12. These file formats can be accessed at saving as shown in Figure 1.13.

1.6 SUMMARY AND OUTLINE OF BOOK

Several textbooks written on the subject of CAD and modelling are available in the market. These are mostly focused on either the engineering concepts or principles of CAD/CAM/CAE in general. On the other hand, the other books that have been published to assist users in operating different CAD systems such as AutoCAD, Solidworks etc. are very specific manuals for whichever CAD system would have been chosen. The novel approach in this book is that it contains a blend or cocktail of four things: principles of CAD and modelling, research and applications of CAD in academia and industry, its use as a manual or tutorial for AutoCAD 2021 and a guide for flexibility in transitioning from one version of AutoCAD to another or transitioning from one CAD system to another. The book is meant for both engineering students and engineering practitioners in a multitude of engineering disciplines. It contains worked examples to assist the user in getting around the principle. The book consists of 14 chapters as follows:

Chapter 1: Introduction

Presents the background to and history of CAD coupled with the requirements for hardware and software, the various software packages available in the market and why AutoCAD was selected as a tool for this book. It also outlines the limitations in the use of CAD to equip readers and users with heads up for effective use and operation. The chapter winds off by looking at the basics for operating AutoCAD 2021 under the Windows environment and the outline of the entire book.

Chapter 2: Two-Dimensional Drawing Primitives for CAD

Outlines the 2D coordinate system and the key building blocks or drawing primitives that are commonly used in 2D drafting such as the different lines and line types, arcs, ellipses, text etc., coupled with guidelines on the various parameters that must be defined prior to the use of such primitives in drawings. Users are also furnished with how to create templates and how these are generally managed commercially or industrially in practice.

Chapter 3: Editing CAD Drawings

Probably the most significant reason for migrating from the drawing board to the computer is the ability to make changes and manage engineering drawings. The chapter provides a comprehensive look at the various editing facilities such as erasing, trimming and extending lines, copying, moving, hatching (sectioning) etc., culminating in a few worked-out examples.

Chapter 4: Dimensioning

The chapter provides wide-ranging guidelines to set out the various types of dimension arrows and styles commonly used in order to automatically dimension models and drawings including tolerances as well as how to modify them.

Chapter 5: Layers

Commercial engineering drawings typically contain a lot of details that often create a clutter of information that can be quite confusing to the user. This is one of the reasons why the biggest paper size A0 is used. This information consists of and can range from finer details of components of machines, architectural details to conform with city by-laws, small electrical components such as diodes and resistors. This chapter provides a guide on how CAD enables the creation of levels or layers on which to place and manage such details. The chapter provides in-depth guidelines on how to manage the layers through locking/unlocking, turning on/off or freezing and thawing them for different purposes.

Chapter 6: Orthographic and Isometric Projections

This chapter contains various worked-out examples on how to lay out the orthographic views (plan, front and sides) in conjunction with title blocks that would have been created. It also outlines how these orthographic views can be used to generate pictorial views in the form of isometric projects using the boxing method, in the same manner as in manual drafting.

Chapter 7: Blocks and Attributes

Another significant reason for migrating from the drawing board to CAD is to avoid repetitive drawing. The chapter outlines the creation of symbols commonly used in engineering drawings such as fasteners, electronic components, hydraulic valves etc., how to store them in CAD libraries and the retrieval process for use in other drawings. The chapter also outlines how non-graphic information such as bills of materials (BOM) can be generated automatically from the block attributes that would have been created with the symbols. A case study is also included to demonstrate the practical applications of blocks and attributes.

Chapter 8: Three-Dimensional Wireframe and Surface Modelling

Chapter 9: Three-Dimensional Solid Modelling

The two chapters outline the different principles employed in modelling objects in wireframe, surface or solid modelling using the 3D surface primitives such as patches using boundary representation (B-Rep) and solid primitives such as cylinders, cones and boxes using CSG. The chapter also provides details on how 2D orthographic views can be generated automatically using CAD and how 3D models are more useful than 2D models in advancing a company's quest for new product development.

Chapter 10: 3D Solid Modelling in Assembly and Analysis

Although almost all the chapters contain exercises and worked-out examples, this chapter contains a specific worked-out example on how to combine all the aspects of 2D and 3D modelling in an assembly and analysis of a machine such as a flywheel, in a typical industrial operation. In the second case study on modelling, a connecting rod for a diesel engine is also included.

Chapter 11: Customization of CAD Software

Most CAD systems like AutoCAD are general-purpose commercial packages that can be used for the production of various types of drawings. This chapter provides information on how to customize the software to suit a user's needs or company operations as well as guidelines on how to insert additional menus and icons to suit their business.

Chapter 12: Management of Models and Drawings for Output

Although printing is no longer very necessary as users can view drawings on site using different gadgets, this chapter outlines the use of model and paper spaces as well as procedures for preparing to print or convert drawings to formats like PDF for ease.

Chapter 13: Further Practical Applications for CAD: Case Studies

This chapter contains three additional case studies intended to enhance users' understanding and practical implications of investing in CAD systems. The three additional case studies cover the use of AutoLISP and EdenLISP in mechanism design and analysis, as well as the integration of the design office and production plant through the linking of graphic (drawings) and non-graphic information (spreadsheets) for production planning and scheduling.

Chapter 14: Typical Examination Questions

The last chapter contains a series of typical examination questions for engineering students pursuing various courses in the broad area of Computer Aided Engineering Design.

Chapter 2

Two-dimensional drawing primitives for CAD

2.1 INTRODUCTION AND ORIGINS OF GRAPHIC DISPLAYS ON COMPUTER

As alluded to in Chapter 1, CAD dates back several decades ago and continues to gain momentum in terms of usage and facilities. Computer graphics (CG) was first developed as a visualization tool, mainly for use by scientists and engineers in the private sector, government and research centres. Because of the increased demands, companies were established to develop specific CAD tools for different applications in many fields of engineering and science. Research institutions and universities also came on board, not only to utilize the CAD systems for training but also to advance and develop other alternatives. This has resulted in a multitude of CAD software, making them more affordable and accessible.

Over the years and in tandem with rapid changes in technology brought on by the Fourth Industrial Revolution, a great deal of specialized hardware and software has been developed for different applications. Most of these devices are driven by CG hardware from the 1960s to date. In general, research in CAD and CG in general cover aspects such as user interface design, visualization, rendering, geometry processing, 3D modelling, simulations and animations, computations and analysis, most of which will be covered in this book and others in the upcoming book on advanced CAD and CAM.

Manual technical and engineering drawings require and consist of several instruments such as straight line rulers, French curves, compass sets, text and other shape stencils. These are necessary to aid in the drawing and development of designs by speeding up the process and achieving accuracy. This was obviously problematic in that some of the required shapes are not regular, hence taking a long time in developing them manually. CADD systems and software have been designed in such a way as to make the development of technical drawings easier and faster through the use of 'building blocks', sometimes commonly referred to as primitives. Before developing any drawings using CAD tools, it is necessary to have a full appreciation of these primitives, their parameters and how they are utilized. This chapter

DOI: 10.1201/9781003288626-2

focusses on the 2D CAD coordinate system and 2D primitives, such as lines, arcs, circles, text etc. that are commonly used in the construction of two-dimensional drawings. The chapter also looks at the origins and manipulation of these primitives using translation and transformation matrices as well as practical applications such as Bezier curves and surfaces.

2.2 TWO-DIMENSIONAL CAD COORDINATE SYSTEM

As shown in Figure 1.7 in Chapter 1, most CAD systems including AutoCAD follow the conventional Cartesian coordinate system (x, y) for display on the graphics screen. Typically, the origin (0, 0) is defaulted to the bottom left corner of the graphics display, although it can be moved and adjusted to any other position. Depending on the units set for the session, whether metric or imperial, incremental values are based on these Cartesian coordinates, y positive going up and x positive going to the right on the graphics display, using the chosen form of dynamic settings, *absolute*, *polar* or *relative* as described in Section 1.5.5 in Chapter 1. In later chapters dealing with 3D modelling, the third coordinate system (z), will be introduced and utilized from that point onwards.

2.3 TWO-DIMENSIONAL GEOMETRIC PRIMITIVES AND PARAMETERS

CAD systems utilize numerous building blocks or primitives to aid in the development of engineering drawings. However, most of them are variations of the basic ones which will be dealt with in this section. While it is true that the use of such primitives speeds up the development of designs and drawings, it is important to note that the correct usage of such building blocks is critical in the sense that the computer and CAD system process what would have been entered, for example, the use of a correct primitive but geometrically placed in a wrong position will invariably result in a wrong output. Accuracy is of utmost importance in CAD graphics, hence the need for users to be extra careful when inputting parameters associated with each primitive.

As generally understood even in manual drawings, there are certain key parameters that must be adequately defined before any building block can be constructed or added onto a new drawing, for example, how long a line required is and where it originates from, centre of a circle and its diameter and orientation of text of any other objects within a drawing. The flexibility of CAD systems is that various other parameters are availed to the user in order to have room to adequately define and thus produce the required object in the correct position and orientation. In addition, these additional parameters are meant to achieve certain complex features in drawings that would ordinarily be difficult to achieve manually or using standard templates. However, this section will focus on the critical/key parameters for each of the commonly used building blocks.

In line with the definitions contained in Section 1.5.5 in Chapter 1, the users must define or set which mode of inputting variables they would have chosen. In AutoCAD, this can be achieved by the command DSETTINGS as explained in Section 1.5.3. This will set the dynamic input appropriate for the input required. However, it is generally advisable to keep the option *Polar* and *Absolute* ON, to avoid mixing coordinates and position of objects.

2.3.1 Line and multiple lines

By far, the most commonly used building block in any drawing is the LINE. Construction of lines follows the principles described in Section 1.5.5 in Chapter 1. Figure 2.1 is an illustration of a drawing that consists of lines and the steps that follow after the figure demonstrate how the drawing is constructed using the three alternatives, *Absolute*, *Polar* and *Relative* coordinates using the command prompt. The same can also be achieved using the mouse and icon menus but for the purposes of clarity, the examples in the book largely make use of the command prompt to enable the user to appreciate the various commands and command aliases that are used in AutoCAD. This makes it easier to migrate from one version of the software to the other.

The starting point always has to be selected by the user, either by selecting a point on the screen using the mouse or specifying absolute coordinates, for example, in the case of Figure 2.1, an arbitrary point (2,2) was chosen.

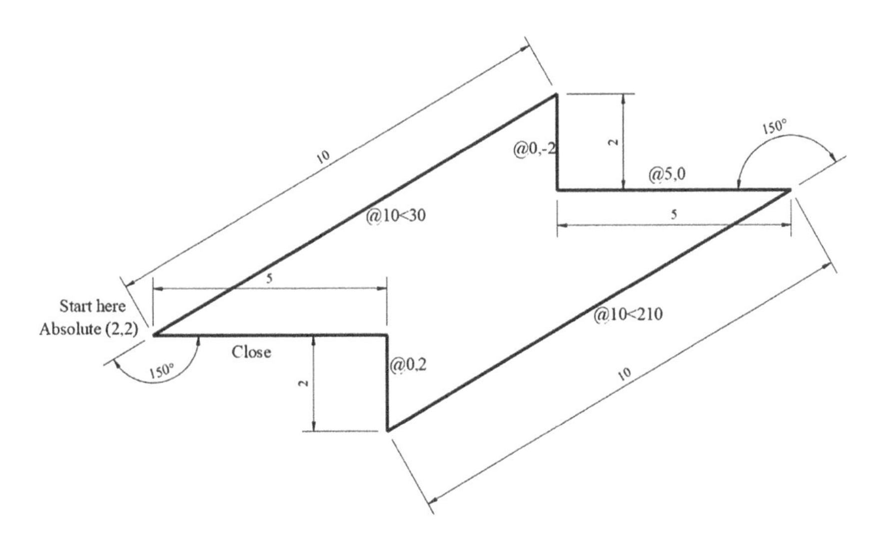

Figure 2.1 Constructing lines using absolute, polar and relative coordinates.

The line drawing is then constructed from that point in any direction. In this case, it was constructed in a clockwise direction as follows:

Command: **Line** or simply the command alias **L** ↲
From point: **2,2** ↲ (absolute)
To point: **@10<30** ↲ (polar)
To point: **@0,-2** ↲ (relative)
To point: **@5,0** ↲ (relative)
To point: **@10<210** ↲ (polar)
To point: **@0,2** ↲ (relative)
To point: **Close** or alias **C** to terminate the line construction

The key parameters for constructing a line are – starting point, length of line and orientation.

Although it is not practised in common, except for regular but complex drawings, there are instances where drawings consist of multiple lines running parallel to each and separated by a given distance. In the earlier versions of AutoCAD, this facility was confined to double lines in which case the user simple specified the starting point, distance between the two lines and orientation of the double line. However, as new versions of the software were developed, multiple lines were introduced and in addition to the parameters similar to double lines, the user is also required to specify the number of lines running parallel to each other.

The default provided with AutoCAD 2021 is the standard double line, 20 units apart. Before commanding the system to draw any multiple lines, it is important to set the distance between the lines as well as set the number of parallel lines required. The Multi-Line (ML) command can be used for this and it allows the user to specify three key parameters – Justification, Scale and Type. Justification refers to whether the reference point or the point where the line is based is, at the Top, Zero (middle) or Bottom whereas the Scale refers to the distance between the two parallel lines. The last option, Style is used to load any particular style that would have been created for that session. Additional parallel lines can be added by first defining a style using the Multi-Line Style (MLSTYLE) command and specifying the offset distance in the second dialogue box that appears after naming the new multi-line style. The use of multiple lines is particularly useful in cases such as architectural drawings where walls are drawn as parallel lines, typically at 230 mm apart (outer walls) and 115 mm apart (inner walls).

The editing of multiple lines, especially at junctions is discussed in more detail in Section 3.2.8 in the next chapter. Figure 2.2 shows typical walls of a simple architectural plan drawn using multiple lines with a scale of 230 mm for the outer walls and 115 mm for the inner wall. There are several line types available for use in AutoCAD, the default one being continuous with fixed width. The rest, which includes dashed (hidden) lines, centre lines, etc.

Figure 2.2 Architectural sketch using multiple lines.

can be loaded by selecting the Line Type Control slide menu, then Other and the rest of the available are visible and can be loaded. However, when such lines are added to a drawing, they may actually not appear as required (dashed or centre etc.). This can be controlled by adjusting the Line Type Scale (LTSCALE), up or down depending on how a user wishes to view them.

2.3.2 Circle, arc and ellipse

The other commonly used drawing primitives are the circle, arch and ellipse. Although there are many ways in which these three can be constructed, this section will focus on the key parameters. Figure 2.3 shows the three building blocks and the key parameters that are required. These are specified through the command prompt by using either the full command or (alias), in this case, CIRCLE (C), ARC (A) and ELLIPSE (EL).

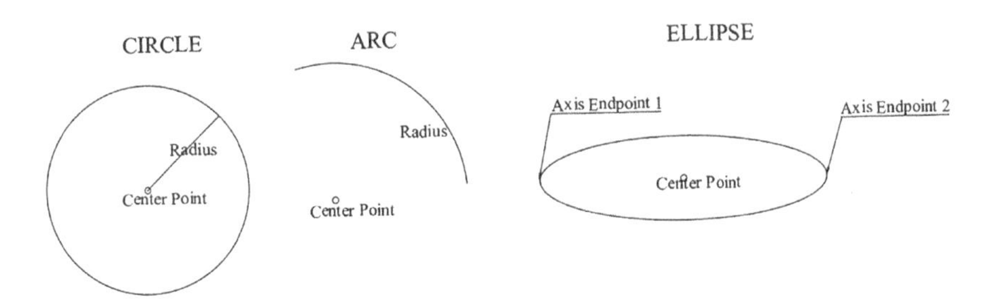

Figure 2.3 Key parameters required for circle, arc and ellipse.

Other than the option for Centre and Radius, the Arc has various other options where users can also specify a centre, coupled with the start and end of the arc. Equally, the arc can also be created by defining three points on it. However, the latter two are less commonly used as these require some form of precision to know the absolute coordinates of the three points. For the ellipse, the three commonly defined parameters are the centre point and the axis end-points of the ellipse as shown in Figure 2.3. These can also be located using the mouse and with the aid of visualization to determine the precise size.

2.3.3 Polygon

To speed up the production of drawings using CAD and to avoid unnecessary and repeated drawing of several lines for regular shapes such as polygons, AutoCAD provides a facility for producing such shapes by simply defining the number of sides and location of the polygon. Figure 2.4 shows an illustration of the construction of polygons in AutoCAD. The key parameters required are – the number of sides and the centre of the polygon. The third key parameter is the radius of the circle that either inscribes or circumscribes the polygon as demonstrated in Figure 2.4, where a heptagon (7 sides) was drawn with a radius of 100, inscribed on the left and circumscribed on the right.

2.3.4 Polyline

As alluded to earlier, the Line is probably the most commonly used building block for any drawing. However, this is provided with a standard thickness (width) as shown in Figures 2.2, 2.3 and 2.4. Invariably, this can be quite confusing especially in a complex drawing, coupled with other lines such as dimensions. Under normal circumstances, in manual drawings, the outer lines or those that constitute the actual drawing, are required to be drawn thicker (bold) than the rest of the lines on the drawing. This enhances clarity and more options will be discussed in Chapter 5, on how layers can be used to reduce the complexity of typical engineering drawings.

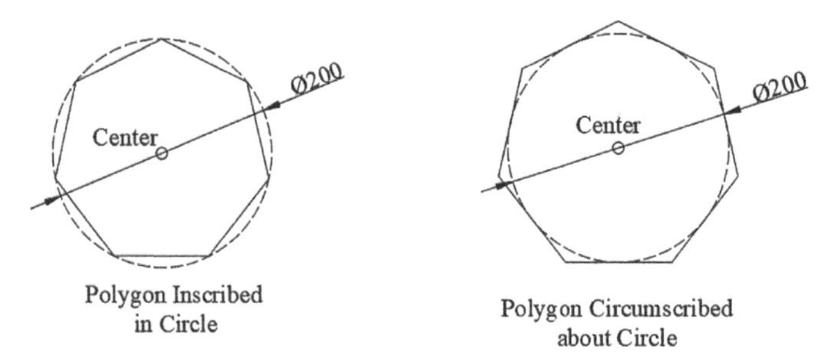

Figure 2.4 Construction of polygons.

The use of thicker lines that can easily be distinguished from other lines such as dimension lines is clearly demonstrated in Figure 2.1. This was accomplished by using Polylines, which are pretty much the same as the standard lines except that their thickness (width) can be adjusted to suit the user's needs. Polylines can be visualized as standard lines stacked together. The ability to adjust them to any thickness implies that the user can actually have several polylines in the same drawing but with varying thicknesses.

The procedure for constructing polylines follows that of line, in that parameters such as the starting point and the next point, defined by *Absolute*, *Polar* or *Relative* coordinates are required. However, immediately after issuing the command PolyLine (PL), the user must define the width (thickness) of the polyline, which can be the same or different from start to end.

2.4 INSERTION OF TEXT IN AUTOCAD DRAWINGS

Except for dimension text, which will be dealt with in Chapter 4, AutoCAD provides a facility for adding text to drawings, such as labels and descriptions of parts of the drawings. For neatness, in manual drawings, stencils were utilized. These were however cumbersome and limited to whatever style was available. Just like any other word processing software package, AutoCAD comes with various built-in fonts at the user's disposal. This helps the user to distinguish the embedded text by way of different font styles, sizes and colours. By default, when a new session of AutoCAD is started, the standard font style is loaded. Any other font styles can be loaded for the session by invoking the STYLE command to get the dialogue box shown in Figure 2.5 from where the user can select and set the required font style.

There are other options that can be set such as the height and angle of orientation of the text. However, due to the fact that these are variable in any given drawing, it is advisable to leave these out at this stage but set them when the TEXT command is invoked to insert any text in a drawing. At any stage of the development of a drawing, text can be added. Using the illustration in Figure 2.5, through the STYLE command, a typical font style such as Times New Roman or any other can be selected and set. Using the TEXT command, three key parameters can be additionally set. These are – Starting point of the text, Height and Angle of inclination to the horizontal, taken anticlockwise (Normal text and writing will thus have an angle of inclination of zero (0), which is the default. Figure 2.6 shows various text sizes and fonts at different orientations.

The other parameter that can be set prior to inserting text into a drawing is Justify, which carries the same interpretation and use as in word processing, commonly applied with Left, Center or Right justifications. Figure 2.6 also shows a typical box that contains text to be inserted, illustrating the key parameters, described above.

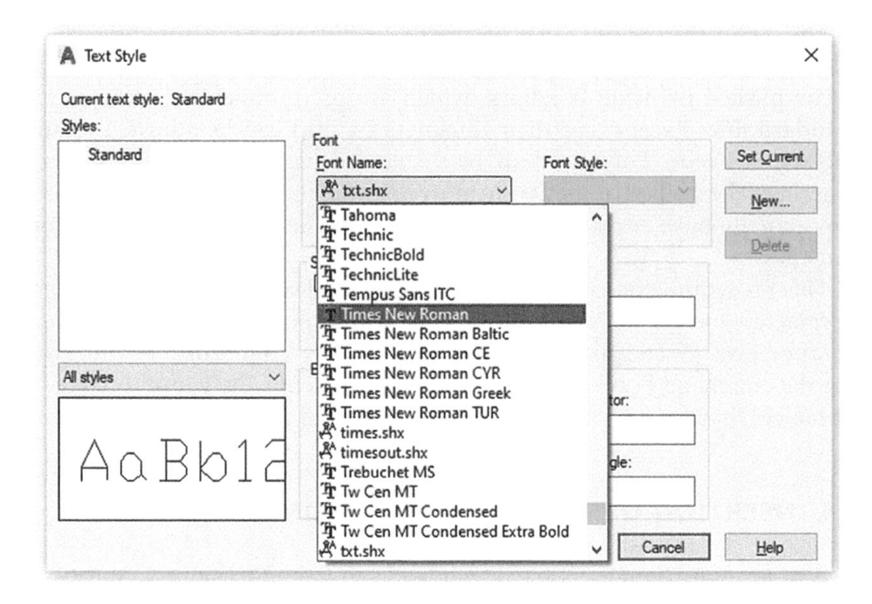

Figure 2.5 Text style dialogue box.

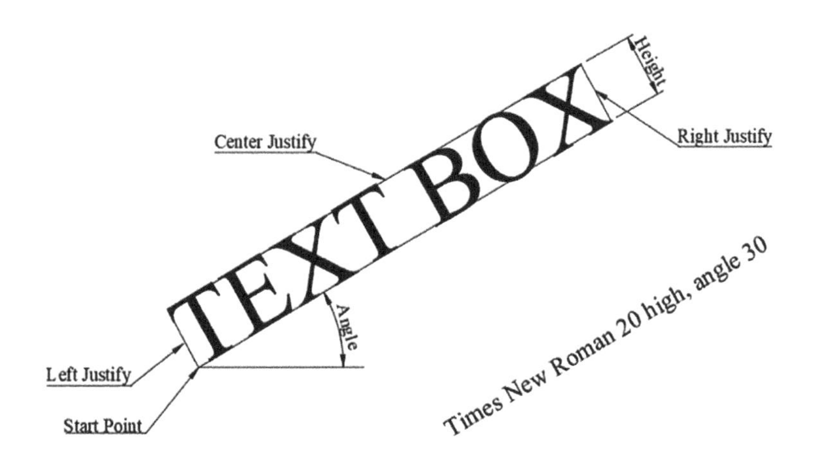

Figure 2.6 Text sizes and fonts at different inclinations.

2.5 CREATING A DRAWING TEMPLATE

As with manual drawings, companies stock drawing papers with partially completed title blocks where users simply insert their drawings and complete the blank spaces on the title block. Different companies have different drawing sheet layouts. AutoCAD provides a facility where such templates can be created and stored as drawing templates (DWTs) for use in future drawings, pretty much the same as blank drawing paper with title blocks in a filing cabinet. Creating one's own DWT customized to their operations or needs is the first step towards customizing AutoCAD for a company's or individual's needs. These can be created to match the different paper sizes as described in Section 1.5.6 and Table 1.2 in Chapter 1.

Figure 2.7 shows a typical Title Block for an Engineering institution that was created for an A3 paper (420 × 297) using polylines and text as previously discussed. Corporate title blocks can be created in the same manner. Logos for the institution or company can also be inserted onto the title block as shown in Figure 2.7, by using IMAGEATTACH command, discussed in detail in the next chapter. Having created the title block, it must be saved as a DWT instead of an ordinary drawing (DWG). It is also important to identify a suitable folder in which this can be stored. By default, AutoCAD stores all DWTs in the Templates folder. However, users are free to select a suitable folder of their choice. Different size templates can thus be created in the same manner and stored. In the event of starting a new session in which a particular template is required, the pull-down menu File is used and then after clicking New, the

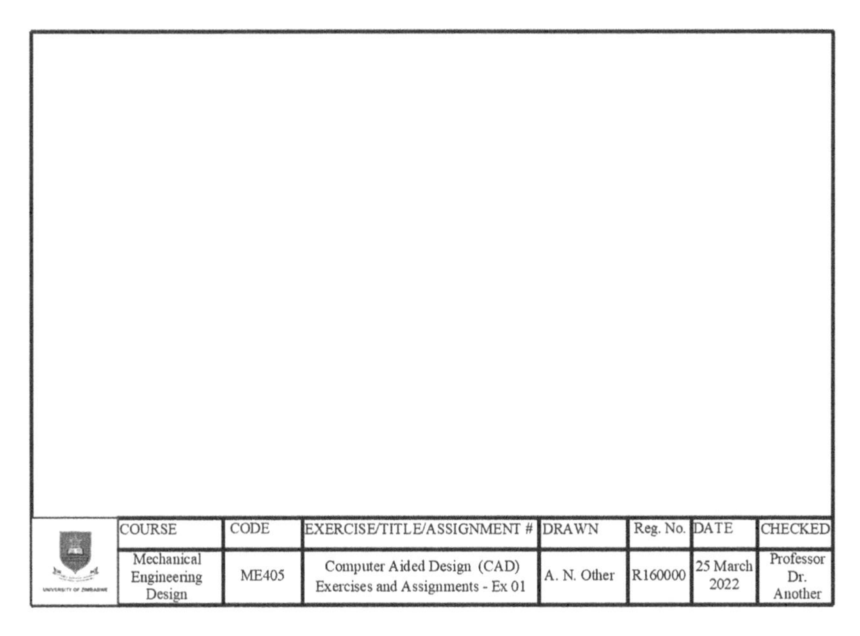

Figure 2.7 Typical title block for an engineering institution.

available templates are then displayed on the Select Template dialogue box that comes out. Once the required template is selected, it is then loaded onto the current AutoCAD session. However, it should be noted that the loaded template is not a saved drawing. At this stage, the required drawings can be developed and inserted over the opened template. Additionally, details for the Title Block at the bottom can also be inserted in line with the drawing in the graphics area. As the drawing progresses, it is advisable to periodically save it with an appropriate name in an appropriate folder.

2.6 PRACTICAL APPLICATIONS

The various building blocks and their parameters described in this chapter are not only useful for the development and construction of engineering drawings but are also the source for practical applications such as linear transformations and translations, curve approximations as well as Bezier curves, based on mathematical trigonometry and theory. Linear matrices are the basis on which these primitives are developed to be ideally located from one point to the other.

2.6.1 Linear transformations and curve approximations

Any graphics system should be capable of performing graphic transformations on existing data. This is made up of three basic operations – rotation, scaling and translation.

Rotation

The object is rotated through an axis perpendicular to that plane as given in Equations (2.1) and (2.2).

$$TP = P' \tag{2.1}$$

$$\begin{bmatrix} \cos A & \sin A \\ -\sin A & \cos A \end{bmatrix} \begin{bmatrix} X \\ Y \end{bmatrix} = \begin{bmatrix} X' \\ Y' \end{bmatrix} \tag{2.2}$$

Where A is the angle of rotation, T is the transformation matrix, P is the original point and P' is the new point after rotation.

Translation

This involves moving the object from one position to another as shown in Equation (2.3). The orientation of the data remains unchanged before and after translation.

$$[X' \; Y'] = [X + TX \quad Y + TY] \tag{2.3}$$

Scaling

As shown in Equation (2.4), scaling involves changing the size of an object which will be done on any of the axes. The scaling factor has to be specified on each axis.

$$\begin{bmatrix} X' & Y' \end{bmatrix} = \begin{bmatrix} X.SX & Y.SY \end{bmatrix} \tag{2.4}$$

Transformations are useful for the designer to view the object in different orientations or positions and also for magnification and diminution.

$$\text{In general,} X_T = XT$$

Where X_T is the transformed vector, X is a vector in the data file and T is the transformation matrix. The coordinate values in X are relative to a reference point X_r, thus Equation (2.5)

$$XT = Tx_r \tag{2.5}$$

For Example: In order to rotate a point $[X\ Y]$ through an angle θ about an arbitrary point at a distance $[R_x\ R_y]$ from the origin, first it will be translated to make $[R_x\ R_y]$ the origin O, thus, Equation (2.6)

$$\begin{bmatrix} X'Y' \end{bmatrix} = \begin{bmatrix} X\,Y \end{bmatrix} - \begin{bmatrix} R_x\,R_y \end{bmatrix} \tag{2.6}$$

Secondly, the required point is thus represented by Equation (2.7) and finally translated again to the old origin as given in Equation (2.8) with the resultant transformation in this case shown by Equation (2.9):

$$\begin{bmatrix} X'' \\ Y'' \end{bmatrix} = \begin{bmatrix} \cos\theta & \sin\theta \\ -\sin\theta & \cos\theta \end{bmatrix} \begin{bmatrix} X' \\ Y' \end{bmatrix} \tag{2.7}$$

$$\begin{bmatrix} X'''Y''' \end{bmatrix} = \begin{bmatrix} X''Y'' \end{bmatrix} + \begin{bmatrix} R_x\,R_y \end{bmatrix} \tag{2.8}$$

$$X''' = (X - O)R + O \tag{2.9}$$

2.6.2 Bezier curves and surfaces

Another practical application of the various primitives described in this chapter is Bezier curves, a useful tool for modelling in computer-aided geometric design (CAGD) and CG. Bezier curves are mathematically defined curves used in 2D CG. Generally, a Bezier curve is defined by four points

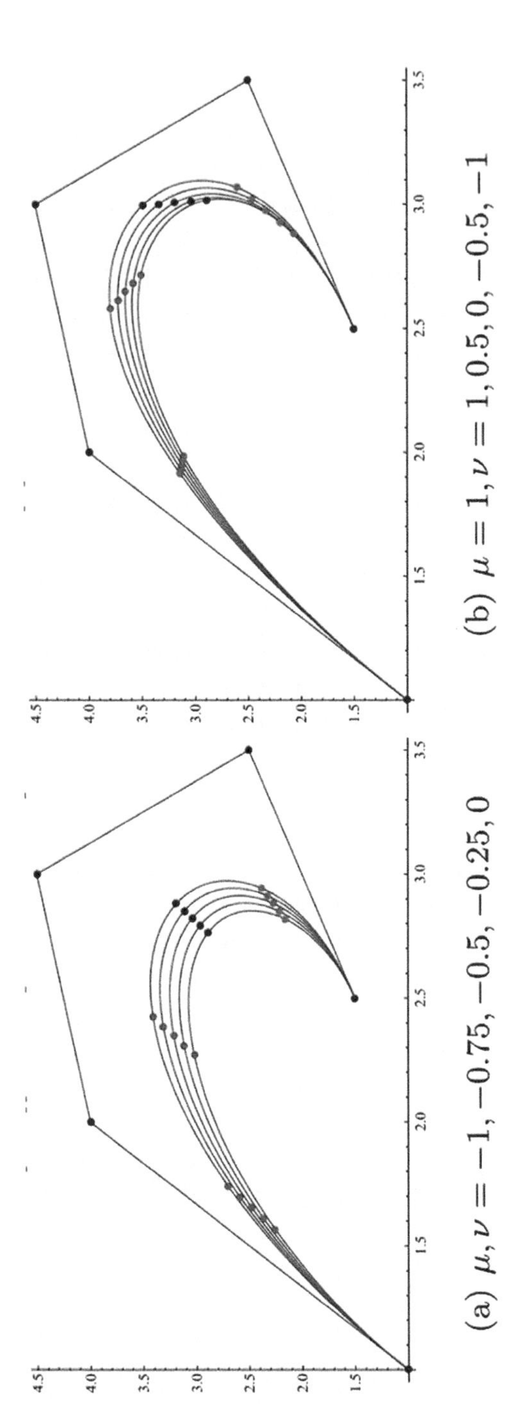

Figure 2.8 Typical Bezier curves with shape parameters.
(*Source*: Maqsood et al., 2020.)

– the initial position and the terminating position (anchors) and two separate middle points (handles) (Maqsood et al., 2020). The latter (handles) are used to alter the shape of Bezier curves. Bezier curves have a wide variety of practical applications dating back to the late 1960s when Pierre Bezier developed the mathematical model for drawing curves in the manufacture of the vehicles at Renault in France.

Following this innovation and invention, Bezier curves have been used for a wide range of applications in engineering, science and technology such as robotics and machine learning, transportation such as railway routes or motorway modelling, animation and simulation, environmental design, artificial neural networks and networks in general (Maqsood et al., 2020). Although they are generally simple to compute mathematically, the original Bezier curves still exhibit some challenges such as noise due to their fixed shapes and positions relative to their control polygons (Chen & Wang, 2003). However, these limitations have been addressed by different researchers over the years, for instance, by introducing shape control parameters and the use of non-polynomial function space as the representation of these types of curves is generally in polynomial form (Han et al., 2008). More details on this will be covered later in the book under surface modelling in Chapters 8 and 9. Figure 2.8 shows typical Bezier curves with shape parameters.

2.7 SUMMARY

This chapter carried on from the introductory chapter and introduced the most commonly used building blocks (primitives) that are employed in the development of technical drawings aided by the computer. The selected primitives described in this chapter are by no means all the building blocks used in drawings but the rest will be introduced in later chapters of the book. The same primitives will be used in the following advanced sections of the book to develop the skills required to generate technical drawings, aided by the computer. The chapter also discussed how these primitives are generated and manipulated using rotation, translation and transformation leading to some practical applications such as the use of Bezier curves in real engineering practices such as motor vehicle body styling and design.

2.8 REVIEW EXERCISES

1. Specifying points and constructing primitives that make up a drawing using CAD can be done through three specific routes – *Absolute*, *Polar* and *Relative* coordinates. Distinguish the three methods, paying particular attention to where each is used more preferentially than the others.

2. There are various ways in which users can input data into AutoCAD such as icon menus, command prompt and pull-down menus. Explain why the use of the command prompt is the preferred option.
3. Many organizations have migrated from manual drawings to Computer Aided Drafting and Design over the years.
   i. What are the major advantages of using CADD over manual generation of technical drawings?
   ii. List and explain some of the limitations and challenges that these companies face in terms of not realizing the benefits of their investment in CADD.
   iii. Out of the many CADD software packages available on the market, list and explain some of the reasons why many organizations prefer AutoCAD.
4. AutoCAD offers various options for users to store their generated drawings. List and explain the most commonly used file formats for such outputs/exports, paying particular attention to where these are used.
5. Describe and explain some of the practical applications of the building blocks used in the development and generation of engineering drawings in CAD.
6. List at least five of the most commonly used building blocks and for each outline the various parameters required to fully define and insert it into a CAD drawing.

Chapter 3

Editing CAD drawings

3.1 INTRODUCTION

So far, the book has covered the most commonly used building blocks and their parameters as well as the basic operating principles for AutoCAD. As with manual drawings, mistakes occur in computer-generated drawings as well and there are instances where users will need to erase mistakes, trim or extend lines, duplicate or copy entities from one position to another and many other changes, as will be covered in this chapter. Obviously carrying out all these tasks manually is not only time-consuming and costly but also requires a lot of experience, precision and accuracy. Changes in drawings are also usually done in order to improve the structure of the drawing, especially for the purposes of presenting this to clients in an acceptable and marketable manner. Such changes can be effected at any stage of the development of a computer-generated drawing whereas with manual drawings if a mistake is noticed towards the end of the drawing, it may be necessary to discard the entire drawing and start afresh. Thus, CAD users need to be equipped with appropriate skills in order to use CAD systems more effectively and beneficially, otherwise the saying 'computers make the generation of technical drawings faster' will not always be true unless the relevant skills are acquired, hence the purpose of this book.

The inherent benefit of migrating from manual drawings to CADD lies in the ability to make changes in the minimal time possible, thus improving productivity and throughput. CAD software packages such as AutoCAD have incorporated numerous user-friendly editing tools and continue to do so in the newer versions of their software. This chapter will not only introduce these basic and advanced editing tools but will also highlight changes made over the years from previous versions of AutoCAD to the one on which this book is mainly based, that is, AutoCAD 2021. This chapter will also provide guidelines on how to manoeuvre and navigate the drawing environment and its surrounding as well as skills to read, view and visualize CAD-generated drawings and models from different perspectives and angles. Such skills are necessary to enhance the user's grasp of the software and its various avenues to generate drawings, not only in AutoCAD but in other

DOI: 10.1201/9781003288626-3

CAD packages as well. It is with such skills that companies can truly realize the benefits of investing in CAD systems, in terms of productivity and company image. A closer look at the practical application of these editing tools will also be done towards the end of the chapter, with a view to outlining some cases where productivity in some organizations has been greatly improved by introducing CAD systems.

3.2 BASIC EDITING TOOLS AND MODIFICATIONS

As with manual drawings, at some point in the development of CAD drawings, there will be a need to make changes such as erasing, moving, copying, etc. In order to achieve this, it is necessary for the users to be familiar with navigating through the AutoCAD environment and its surroundings as shown in Figure 1.7 in Chapter 1. The following sections focus on the different editing tools and those that are closely related will be handled in the same section.

3.2.1 Navigating through the drawing and environment

It is assumed that those who are using this book as a guide to AutoCAD have the software installed to work in conjunction with a standard mouse which has three components – the Left Mouse Button (LMB), the Right Mouse Button (RMB) and a roller in between the two buttons. The mouse can thus be used to navigate through the graphics window (drawing editor) where a cross-hair appears as well as the icon menus that surround the drawing editor, where a pointer appears. The LMB is the most commonly used button for making a selection and the RMB is often used for bringing up the quick properties menu as the drawing progresses, while the roller is used to zoom in (reduce viewing size of drawing but not scale) or zoom out (increase the visible size of the drawing on display). This is all done in conjunction with the snapshot of the AutoCAD screen as shown in Figure 1.7 in Chapter 1.

The size of the cross-hair can be adjusted to the user's preference by selecting the **Tools** pull-down menu, the **Options** and under **Display**, the cross-hair size can be adjusted up or down from the default of 10%. Similarly, the background colour or fonts for the current session of the drawing editor and its surroundings can also be adjusted to the user's preference by selecting the same **Tools** pull-down menu, **Options** and **Display** and then selecting **Colour** or **Fonts** to adjust accordingly. In the previous versions of AutoCAD, this was done under the **Tools** pull-down menu but selecting the last option of **Preferences** was then used to arrive at the same dialogue box for **Options** (newer versions) or **Preferences** (older versions).

Generally, the default AutoCAD screen is as shown in Figure 1.7 in Chapter 1, with the Drawing Icons on the left panel, the Modify Icons on the right panel and the Pull-Down and Slide Menus at the top. These can be selected and dragged to any position that users prefer. There are numerous other menus that can be added such as the UCS display, 3D primitives etc., and this can be done by right-clicking on the black space just above the drawing editor and then selecting AutoCAD where the several menus can be turned on or off by ticking or unticking them.

As a drawing under development progresses, it becomes more complex as more entities are added, making visibility poor. In order to improve visibility, two things can be done, that is, zooming in and out using the mouse roller or even more importantly panning the drawing from one position to another using the Pan Realtime icon (open palm), found among the slide icons just above the drawing editor. Once the panning is complete, the RMB can be used to terminate the process before continuing with the drawing. Several other icons or slide menus and manoeuvres within and around the drawing environment will be dealt with in the respective chapters that will follow.

3.2.2 Entity properties

All building blocks (primitives) described in Chapter 2 have their properties and parameters attached to them. These include the type of lines, height, angle and style of text, etc. These make up the entity properties for all objects that are inserted in a drawing. At any given time, any of these entities can be modified, for instance, text height or its angle of inclination. There are two ways in which this can be done, both by first selecting the entity and then right-clicking (RMB) to bring out the Edit menu and then either selecting **Quick Properties** or **Properties**. Quick Properties provide the basic properties of the entity selected, for example, for a line, quick properties include colour, layer and length, for a polyline, it includes the same properties as the line but will also include the line width (thickness), for text, quick properties include layer, style, height, style and angle.

Selecting the option Properties from the Edit menu will provide all the properties of the selected entity including the basic ones, for example, for a line, in addition to the basic entity properties mentioned above, additional properties will include, the (x, y, z) coordinates for the start and end of the line. Likewise, several other properties will be availed for the other building blocks if the full properties are selected.

At any given time during the development and construction of a drawing, these entity properties can be changed by either of the two options and the entity will accordingly change. It must also be noted that several entities' properties can be changed at the same time by selecting the entities required and then adjusting the properties. However, it should also be noted that, in the event that two or more different entities are selected, the properties that can be adjusted are only those that will be common to the two or more entities.

AutoCAD also provides a facility to match properties of an entity in the drawing to another one. For example, if a user wishes to change the width or thickness of a polyline in the drawing, or the height and angle of inclination of text within the drawing etc., to that of similar entities within the same drawing, the command MATCHPPROP or its icon (tip of a fountain pen on the slide menu), then select the reference entity and repeatedly select the entities whose properties need to be matched with the reference entity. It should however be noted that this option is limited to certain properties such as colour, width, height of text, etc. but does not include the start and end points of lines.

While the selection of entities can be done using the LMB, it must be noted that this can be done in three ways. First, hovering the cursor or cross-hair over part of the entity highlights it and thus can be selected. When multiple entities are required to be selected, this can be done by enclosing them in window but it is important to understand the difference between opening the window from left to right or from right to left. Second, in order to select enclosed entities from left to right (continuous window), all the required objects must be enclosed within the selection window whereas selecting from the right to the left (dashed window), only parts of the required objects need to be within the dashed window and will all be selected. This is illustrated in Figure 3.1. There are other circumstances as shall be seen in later sections such as the selection of entities by enclosing them in window should be done in a particular way, either left to right or right to left.

3.2.3 Erase

Probably the most basic of the modification commands is the ability to correct a mistake by using the Erase (E) command, upon which the entities to be erased can be selected using any of the three options described above. Sometimes the user may require to erase all objects on the screen in order to refresh their drawing editor. Instead of selecting objects, ALL can be used to erase everything on the screen or by reloading the drawing template to start afresh.

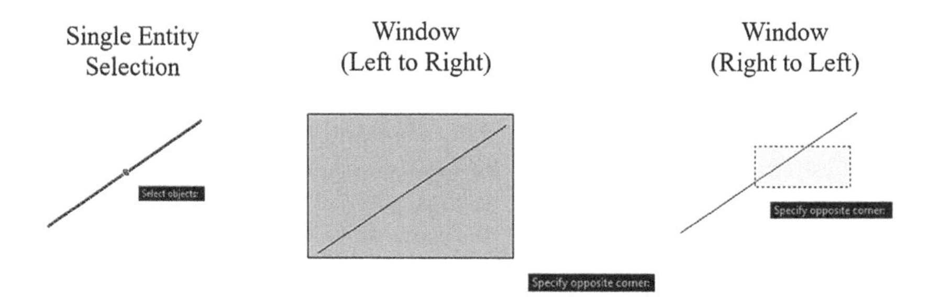

Figure 3.1 Selection of entities in AutoCAD.

3.2.4 Move or copy

In the event that a user realizes a mistake during the construction of a drawing, that a particular entity is in a wrong position or that a duplicate of the same entity is required in another position, the commands Move (M) or Copy (CP) can be used to effect the changes respectively. The two commands have been grouped together because the procedure to effect the change is similar. To effect either of the two changes, the user must define two properties, first, the reference point from where the entities are to be moved from and the new point where they should be moved to.

The reference points can be random or specific but the new point must be defined appropriately, all done by object snap points described in Chapter 1, such as endpoint, centre, midpoint, etc. The difference between Move and Copy is that an entity that will be moved, will not leave a copy or duplicate in the original position and for copying, a duplicate is left in the original position. In addition, the copy, by default will allow the user to make multiple copies of the object from a given reference point to other selected points and this command needs to be terminated (RMB) or Enter before moving on with the construction of the rest of the drawing.

3.2.5 Trim and extend

The proper construction of a technical drawing in CAD entails entities placed in the correct absolute positions. Where a line is supposed to be connected with another, the connected position must be at the absolute value of the endpoint of one line and the beginning of the next line. Quite often, the visibility of these connection points can mislead the user into believing that the two objects are connected. The lack of connection will create challenges for other modification commands such as hatching as will be seen later in this chapter. To ensure that entities are connected at the same absolute (x, y, z) point, objects such as lines can be either trimmed or extended from their original positions to the point of absolute connection.

Trim (TR) implies cutting an entity to an absolute position while **Extend (EX)** implies stretching an object from an original position to an absolute point. In both cases, an edge to where the object must either be trimmed to or extended to must be defined. For trimming, this will be the cutting edge and for extending, that will be the boundary edge. In the earlier versions of AutoCAD, the user was required to select these edges clearly, including the entity that needed to be trimmed or extended.

However, the later versions of AutoCAD have simplified this by simply allowing the user to select the object that needs to be extended and automatically the system highlights the possible edges to which the selected entities can be trimmed or extended to. Figures 3.2 and 3.3 show illustrations of trimming and extending as derived from the earlier versions of AutoCAD.

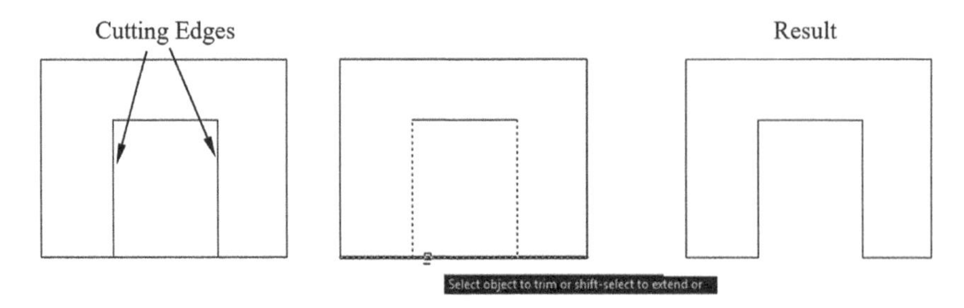

Figure 3.2 Illustration of trimming (earlier versions).

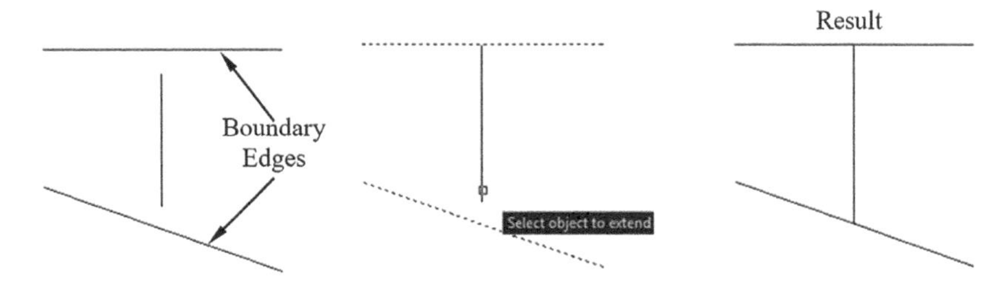

Figure 3.3 Illustration of extending (earlier versions).

3.2.6 Stretch

Quite often users of CAD encounter challenges when they realize that the object they have created falls short of the design requirements. In this case, instead of erasing the object and starting afresh, it may just require stretching it a little such that it conforms to whatever is required. The STRETCH command in AutoCAD is used to stretch the portion of an object partially enclosed by the selection window and then specify the base and direction of stretch. To achieve a proper stretching of the objects, it is recommended to use a right-to-left window selection (dashed) of the entities to be stretched. Figure 3.4 shows an example illustration for stretch. Assuming that a user is

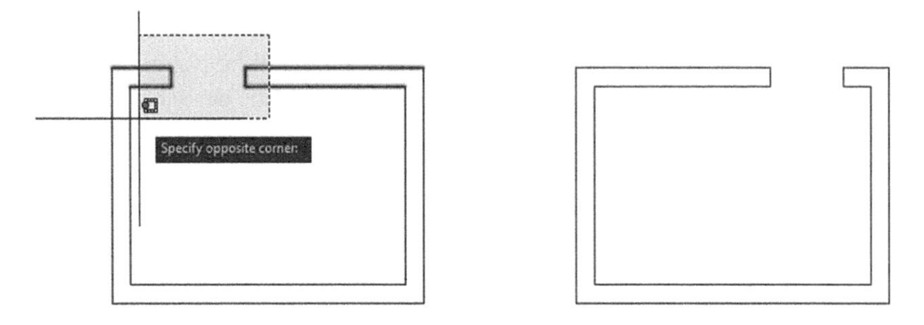

Figure 3.4 Stretching entities to reposition an opening in architecture.

constructing the plan of a building as shown on the left of Figure 3.4, they realize that the opening should have been more to the right. The command STRETCH is used and then a selection of the entities on the opening as shown in Figure 3.4 is done, followed by selecting the base point and then stretching these entities to the right to position the opening of the plan correctly. The stretch command thus avoids the repetitive procedure of having to erase the opening and drawing it correctly in the right place. In case there are several openings that need to be shifted by stretching, Multiple Stretch command (MSTRETCH) is used. The different entities to be stretched can then be selected one by one and then stretched together.

3.2.7 Break

Objects created in AutoCAD such as those described in Chapter 2, that are used as building blocks for any drawing, are drawn as single entities. Quite frequently a user may require these to be multiple entities in order to carry out certain other modifications on them. For example, a straight line may require to be opened up or simply broken into two or more entities in order to utilize the broken points as object snap points.

The Break (BR) command can be used for this purpose, in which case, the point at which the object is selected becomes the first point of the break and as the cursor is moved, the segment of the entity to be removed (broken) is highlighted. An appropriate second point will thus break open the entity. In the event that the user does not wish to open up the gap but simply to break the line, the **Break at Point** icon (from the Modify I) icon menu is used. In this case, the object is first selected and then the point at which the break point is required is selected.

3.2.8 Explode

Almost for a similar purpose as Break, the EXPLODE command can be used to break a single entity such as a rectangle into the four lines that make it up. This command breaks a compound object into its component entities. This is usually important and necessary when only one of the component entities needs to be modified or used as a reference without affecting the rest of the components that make up the compound object. Typical objects that can be exploded are polygons, polylines, blocks, etc., which are mainly composed of many entities but form one compound object. This will be used more in detail, particularly in Chapter 7 on blocks and their attributes.

3.2.9 Scale and zoom

Although visually, zooming in and out of a drawing will decrease or increase the visibility of a drawing, its actual size does not change. If a user wishes to change the absolute measurements of an object, the command Scale (SC) can be used. In this case, the object(s) that need to be scaled up or down is selected, followed by specifying a base point from where the scaling starts and then specifying a scale factor n. If $n > 1$, then the objects will be increased in size, n times and if $n < 0$, the objects will be decreased in size, n times.

In order to improve the visibility of the drawing under development, the Zoom (Z) command can be used and this magnifies the area that is zoomed but does not change the actual measurements of the objects. The Zoom command has various options, the default one being a window selection of the objects that require to be magnified. The other commonly used options include; **Zoom-All** (magnifies all the objects to the limits of the available screen in session), **Zoom-Extents** (magnifies all objects to the limits of the screen that would have been set at the beginning), **Zoom-Scale** (a scale factor can be used to increase or decrease visibility but this should not be mistaken for the Scale command), **Zoom-Previous** (restores the previous display after a zoom) and **Zoom-Dynamic** (allows user to dynamically shift the display to a displayed window that they can interactively adjust).

3.2.10 Editing polylines

As alluded to in Chapter 2, polylines are flexible for producing outlines of drawings because they allow the user to increase the width or thickness of lines, thus distinguishing them from ordinary lines or dimension lines. The command Poly-Line Edit (PEDIT) is used to modify properties of such entities that include the width/thickness. The same command is used to convert an ordinary line into a polyline. One of the options for the PEDIT command is to Join connected polylines to become one. This is particularly useful for later chapters on 3D modelling.

3.2.11 Multiple line editing at junctions

The joints for multiple lines discussed in Chapter 2 can be quite cumbersome to clean up after a construction. AutoCAD offers a facility to clean these with ease through the utilization of the Multiple Line Edit (MLEDIT) command. This command brings up a dialogue box that contains various options for cleaning/trimming the junctions as illustrated in Figure 3.5.

Assuming that a user is in the process of constructing the drawing displayed on the left of Figure 3.5, with the ultimate result required as shown on the right side of the same figure, the various options offered in the Multi-Line dialogue box will enable the cleaning up/trimming of these junctions. The Open Cross option is used to trim the central junction, while the Corner Joint is used to trim the top and bottom left junctions and the Open Tee for the remainder of the junctions.

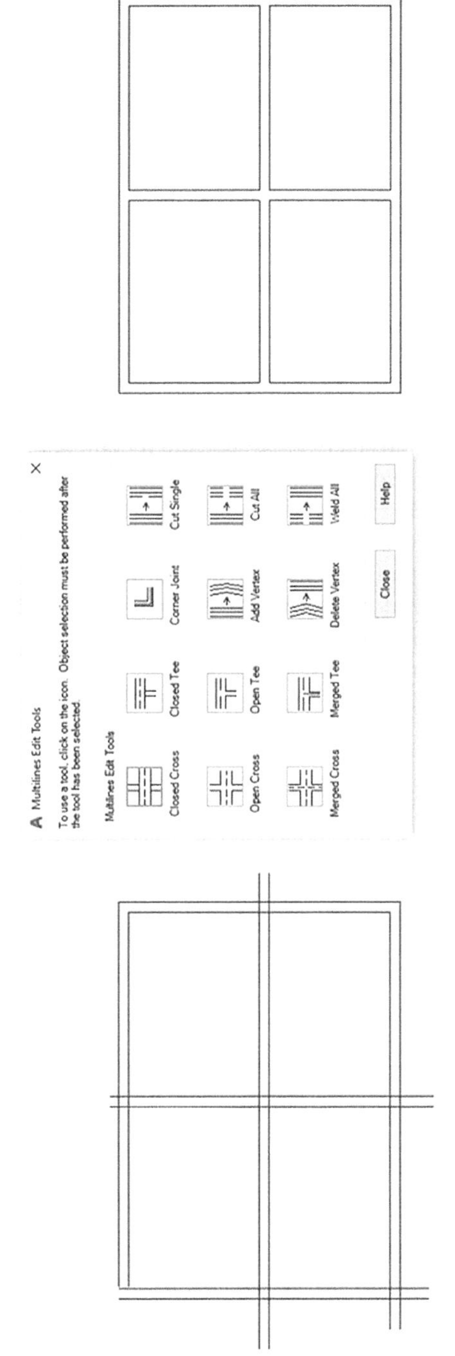

Figure 3.5 Multi-line editing tools.

3.3 ADVANCED EDITING

The previous section focused on the basic editing tools that are used in the development and construction of technical drawings using CAD. This section will focus on the more advanced editing tools, some of which are combinations of the previous basic tools.

3.3.1 Rectangular and polar arrays

The Copy command which was used in the previous section on basic editing commands can be used to duplicate multiple objects within a drawing to other places within it. However, if the duplicates are required in a particular order, especially in an array arranged either in a rectangular format (rows and columns) or dotted around a central point (circular), then rectangular and polar arrays are used to accomplish such a task, the Array (AR) command. For the rectangular array, three parameters must be defined: first, the object that needs to be duplicated, second, the number of columns and third, the number of rows. For the polar array, three parameters must also be defined: first, the object to be duplicated, second, the centre of the array and third, the number of items around the central point. Figure 3.6 illustrates both the rectangular and polar arrays, showing the parameters that must be defined for both. For flexibility, there are several other options under the Array command at the user's disposal that can be explored for a specific arrangement of the objects in either the rectangular or polar array.

3.3.2 Offset

Duplicates of lines, arcs, circles and polylines can also be made from an original entity to another position offset by a required distance. The OFFSET command requires the user to specify three things – the object to be offset,

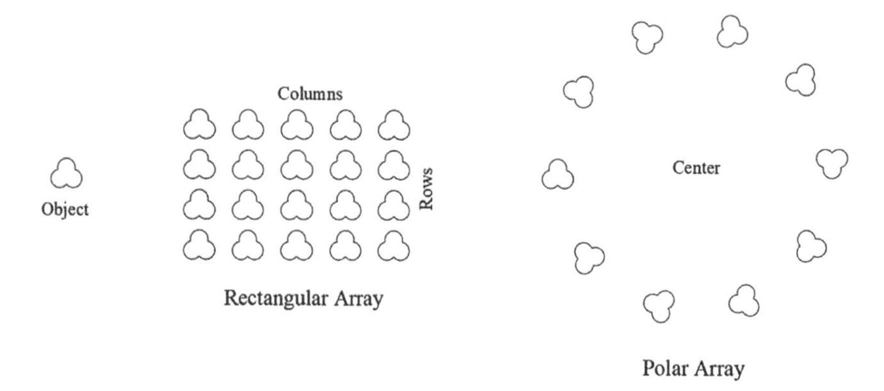

Figure 3.6 Rectangular and polar arrays.

the distance through which the object should be offset and the direction from the original object to where the object must be offset to. This is particularly useful in constructing architectural walls that have variable widths but can equally be done using multiple lines. The choice of which option to use will depend on what is being constructed. Whichever option is shorter or easier for the user should be pursued.

3.3.3 Fillet and chamfer

Instead of drawing arcs and lines to produce fillets and chamfers/bevels where two lines meet, then followed by trimming etc., AutoCAD offers the facility for automatically generating these modifications. This significantly reduces the time to carry out the tasks compared to creating circles and lines and then trimming them.

For the fillet, the user needs only to define the fillet radius followed by the two edges to be filleted while for the chamfer, the first chamfer distance (distance from where the two edges meet to the point where the bevel starts) for the first edge and the second chamfer distance, defined in the same manner on the second edge. Figure 3.7 shows illustrations of the fillet and chamfer on an object obtained using the FILLET and Chamfer (CHA) commands.

3.3.4 Mirror

The Mirror (MI) commands also allows users to create a duplicate of objects or entities laterally inverted (mirrored) about a chosen axis. Three parameters must be defined in order to achieve this and these are – the object(s) to be mirrored, first and second points of the mirror axis.

The command also offers the option to either retain the original entities or discard them and simply remain with the mirrored objects. Figure 3.8 shows an illustration of a mirrored object with the required parameters.

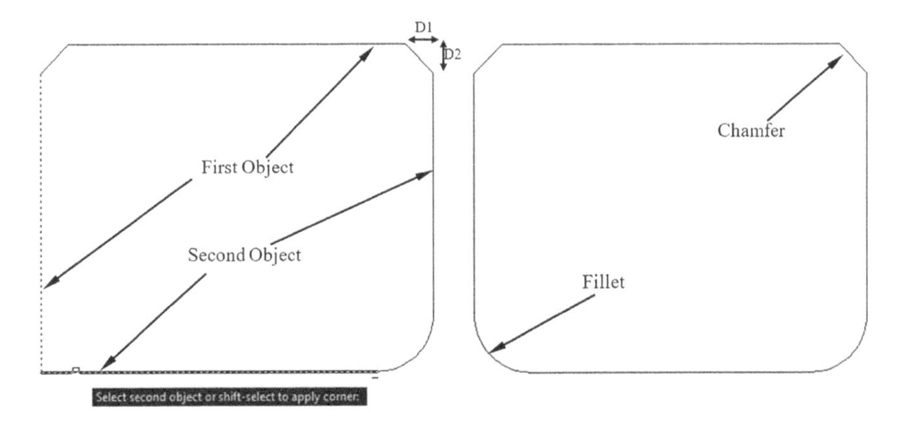

Figure 3.7 Fillet and chamfer.

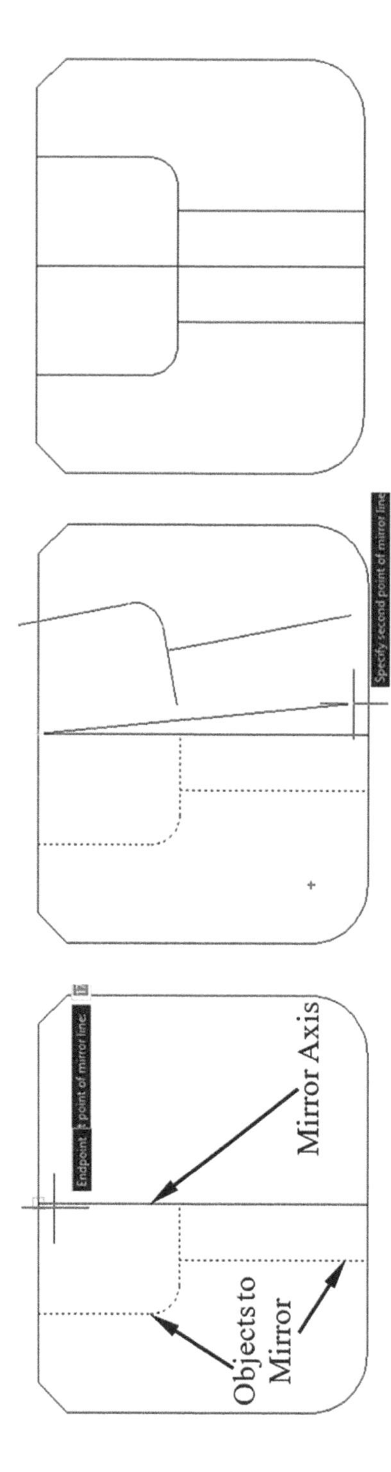

Figure 3.8 Mirrored object.

3.3.5 Divide, measure and distance

Following on from the Break command that was discussed earlier, the Divide (DIV) command allows users to split an entity such as a line, circle or arc into an equal number of segments. Thus, the command requires the user to select an object and specify the number of divisions required. However, this does not split the entity into segments as covered under the Break command but it allows users to insert markers that can later be used as nodes to help in identifying exact locations for breaking such entities.

After dividing the object, nothing actually shows that it has been divided until these pointers or nodes are inserted. These can be inserted using the **Format** pull-down menu and then **Point Style** from where a dialogue box displays several available point styles to choose from. Alternatively, the PTYPE command can be used for the same. The sizes of the point styles can also be adjusted using the same dialogue box. In the event that a user wishes to break the object at the node points, the Break point icon from the **Modify I** menu is used and then the object is selected followed by specifying the node from the object snap points.

After the object has been broken at the various node points, the pointers can be removed to reduce the drawing's complexity. The MEASURE command does almost the same as the Divide command except that it creates point objects or blocks at measured intervals along a selected entity such as line, circle or arc. The resulting points are always located on the selected object and their orientation is determined by the XY plane of the UCS. The difference with the Divide command is that the pointers are placed at specific intervals that the user would have chosen but what this also implies is that the last segment may not be equal to the other segments because of the specific figure that would have been set.

The MEASURE command should not be confused with the Distance (DIST) command which is used to find the distance between two points, provided in all three directions (x, y, z) as well as the direct distance measured between the two points. Figure 3.9 shows an illustration of the Divide and Measure points on a segment.

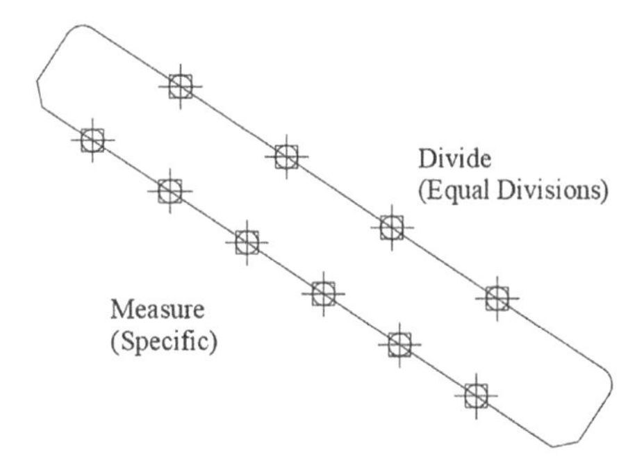

Figure 3.9 Divide and measure on segments of a drawing.

3.3.6 Hatching

Whether it is architectural, mechanical or electronic drawings being developed, the more details there are, the more complex the drawing can be, especially if it is an assembly comprising several components. Even with manual drawings, these have to be opened up for clarity using the technical drawing conventions for sectioning. Most CAD packages refer to sectioning as hatching. Where there are many different parts in a section, the styles used to differentiate them range from lines drawn at an angle, covering the entire section of the component that is hatched.

While these styles are limited in manual drawings, AutoCAD offers a variety of these hatch styles and the user can also define their own, as will be seen in Chapter 11 on customization. The key parameters that the user must define consist of the boundary area in which the hatch pattern must be inserted, the hatch pattern and the scale (dense or spaced patterns). The area to be hatched must be adequately enclosed within a boundary whose edges must be properly connected otherwise if they are not, the system will either not hatch or will hatch incorrectly.

The dialogue box to set all the parameters is loaded by using the HATCH or Boundary Hatch (BHATCH) commands. Pick points are used to select the region within which the hatch pattern falls. The pattern can be selected from various patterns available, while the angle of orientation of the pattern and scale can likewise be set. Before accepting the settings, the dialogue box also offers the option to **Preview** the resultant hatch. If it is not satisfactory, the Esc button is used to return to the dialogue box to adjust the settings until the preview is satisfactory. In older versions of AutoCAD, unlike the HATCH command, the BHATCH command helps the user to better use as it supports boundary hatching, allowing the user to pick a point that is adjacent to the boundary they want, and this command lets AutoCAD automatically search for the nearest entity, then constructs a closed boundary by tracing in a counterclockwise fashion to look for intersection points as well as connecting lines or arcs. Figure 3.10 illustrates an object that was hatched in two regions, showing the parameters that are set to achieve this.

3.3.7 Regeneration

Quite often as designers develop their drawings, there is a lot of Zooming in and out of the drawing. Sometimes when the zooming is excessive, where objects are made much smaller than they are supposed to be, entities such as circles may be distorted such that they appear as polygons (series of lines making up the small) when the user returns to the original view. In this case, the REGEN command is used to restore the smooth circles instead of the polygon. From time to time as the drawing is in progress, a lot of unnecessary items are deleted but sometimes they do not disappear completely but will make the drawing unnecessarily big in terms of storage. The REGEN

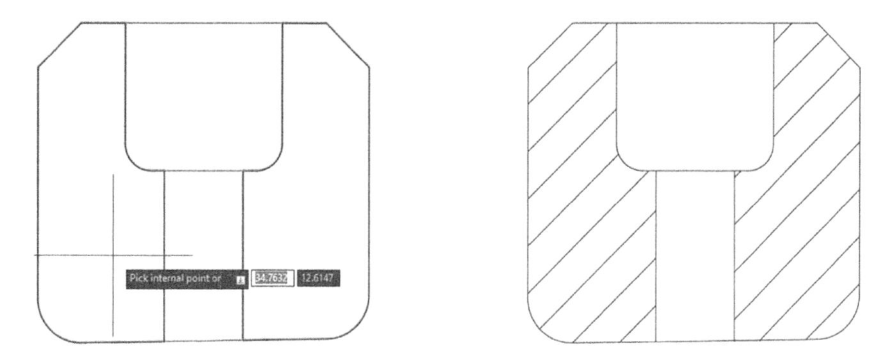

Figure 3.10 Illustration of hatched sections of an object.

command can also be used to get rid of these unnecessary items. In the earlier versions of AutoCAD, the PURGE command was used to perform the same task. In the newer versions, the PURGE command can also equally be used to get rid of unused named objects from the drawing and these can be chosen from a list that appears on the dialogue box.

3.4 APPLICATIONS AND PRODUCTIVITY

The various modification parameters outlined in this chapter are probably the most significant advantage of investing in CAD systems compared to manual drawings. As drawing office personnel – engineers, CAD managers, draftsmen, etc. – CAD systems can be used widely in a variety of ways, from sketching to detailed design, bills of materials as will be seen in Chapter 7, through to production of drawings for clients. These processes can be time-consuming, and in bid to beat global competition, organizations have to find ways to increase throughput and productivity in the shortest time possible, and what better way to be able to quickly make changes and modifications to satisfy customers. These modifications are part of CAD's contribution to positively impact productivity by eliminating inconsistencies, aiding standards conformance, troubleshooting challenges, and in so doing avoiding errors, budget loss and risk (Hirz et al., 2011).

Unless a company has invested a lot and stocked on manual drawing paper, the creation of templates significantly reduces the amount of time spent in recreating them every time a new drawing is generated. Drawing templates store a variety of settings such as drawing-specific variables, layers, line types, plotter settings, symbols, sheet borders and other data. The ability to copy, duplicate or move objects also reduces time in comparison to manual drawings, even more, when multiple copies can be made and arranged in defined arrays. Where machine, architectural or other complex drawings are being developed, these can easily be managed and better analysed using facilities such as mirror, hatching etc., which would otherwise take longer to just produce those strands of section lines one by one. With the advent of the Fourth Industrial Revolution, any drawings under development can concurrently be worked by various personnel from design, drafting, detailing and bill of quantities.

3.5 SUMMARY

Carrying on from the building blocks in Chapter 2, this chapter has focused on and discussed the various modification tools that are available in AutoCAD. These ranged from the very basic ones such as erasing, trimming, extending and breaking duplicating through to the more advanced option, some of which are a variation of the basic ones. These included arrays of objects arranged in a particular manner, creating objects laterally inverted, dividing and measuring in contrast to getting the actual distance from one point to another.

AutoCAD has also introduced ways in which objects can be shaped quickly without having to go through even some of the basic commands such as trimming and extending, by making use of fillets and chamfers to quickly and automatically shape objects to the desired configuration. Probably one of the most important facilities includes the ability to hatch sectioned objects by simply defining the boundary within which the hatch patterns are to be applied. The chapter culminated in analysing how the various modification can contribute to the productivity and throughput of drawings. With the basic skills for generating building blocks and the ability to manipulate and modify, the next few chapters will look at their applications and additional techniques to enhance the skills in order to realize the full benefits of investing in CAD.

3.6 REVIEW EXERCISES

1. Figure 3.11 is an illustration of the front elevation of a building demonstrating the practical application of hatch patterns in architectural drawings. Construct the elevation in AutoCAD and use appropriate hatch styles and scales to match the figure.
2. Using illustrations, outline the similarities and differences between the two pairs of modification commands as employed in AutoCAD – *Move* and *Copy*, *Extend* and *Trim*.
3. What are the basic entity properties for the following CAD primitives: Line, Polyline, Circle, Text, and how are these properties modified in the course of developing a drawing using AutoCAD?

Figure 3.11 Illustration of the practical application of hatching in architecture.

4. What modification command in AutoCAD can be used to break a compound object/entity that consists of several other entities? Explain where this option is useful in the development of drawings using CAD.
5. Although AutoCAD has various types of lines available within the system, only the continuous line of standard thickness is loaded at the beginning of each session:
   i. Explain how you can load the rest of the line types.
   ii. After constructing a drawing that utilized one of these line types, the dashed (hidden) line, the display still consists of the continuous line. Explain what you need to do to improve the visibility such that the hidden line will appear.
6. List 10 reasons based on the various AutoCAD modification tools why it is essential for companies to invest in CAD systems, focusing on their productivity and throughput compared to manual drawing systems.

Chapter 4

Dimensioning CAD drawings

4.1 INTRODUCTION

Unless engineering designers and draftsmen are simply sketching or drafting an idea at the brainstorming stage or conceptual stage of the design process, all technical drawings are by nature developed knowing fully well the sizes and magnitudes of their components and the overall size of the finished product. As such, once a drawing has been made, it is necessary to dimension it before a hard copy is produced or before the drawing is taken from the drawing office for manufacturing or to the site. Before any job can be carried out, the manufacturing technicians or the contractor on site needs to know how to set the component for manufacturing or set the site for construction, derived from the dimensions provided by the detailers or designers. Simply putting the scale on the drawing and expecting the technicians or contractors to estimate or measure from the hard copy is tantamount to errors and thus should never be practised as is unconventional to take measurements from a hard copy. All technical drawings for use in production, laboratory or construction site must be sufficiently dimensioned to ensure clarity while at the same time, the provided dimensions should not be superfluous, otherwise, it creates unnecessary confusion and clutter on the drawings.

Regardless of whether the dimensioning is being carried out manually on the drawing board or aided by the computer such as using AutoCAD, detailers or designers are bound to present such drawings with clarity such that it should never be necessary to refer back unless it is meant for suggesting improvements. Depending on the country or region, there are various drawing standards that are supposed to be adhered to in such a way that whatever the detailers produce should be clearly understood by manufacturers in production or contractors on site. These standards are inclusive of dimensioning, whether manually carried or computer-assisted. In addition, the units used in such standards are also adopted and derived from the same standards in such a way that even when a dimension is missing and has to be estimated, the users will know whether it is imperial (inches/feet) or metric (millimetres). This book is based on the SI units and any measurements

DOI: 10.1201/9781003288626-4

provided but not specified must always be assumed to be in millimetres (mm). The purpose of this chapter is to provide guidelines and skills to efficiently dimension CAD drawings using AutoCAD using the various tools embedded within AutoCAD based on standard drawing conventions and SI units. This is ultimately meant for organizations investing in CAD systems to fully realize the benefits by improving productivity, clarity and throughput as well as return on investment.

The dimensions of components and overall sizes of these put together to make assemblies of machines or architectural buildings are required to be communicated by dimensions inserted on the drawing. Detailers should also be very careful how they specify dimensions, especially in machining where certain precisions cannot be attained but should leave enough room for allowances. In the case of mass production or assemblies in factories, special care and attention must be placed on dimensions of parts that must fit together, in such a way that sufficient tolerances are allowed, whether it will be interference, clearance or transition fits. More details on this will be covered in Chapter 10 on Constructive Solid Geometry (CSG) of an assembly.

In such precision machining, detailers and designers must specify an acceptable range for holes, for instance, and the shafts that fit in them, for example, for a 10 mm shaft, the hole to fit this can be specified as 9.995 mm to 10.005 mm (tolerance of 0.01 mm), depending on what type of fit is required. In some instances, where a very tight fit (interference) is required, heat may be used to force through a press fit. The implications of specifying an improper tolerance in dimensions for parts that fit together is that it will inadvertently result in the final product such as a machine not functioning optimally or not functioning at all, thus reducing the life expectance of the machine. Most CAD packages such as AutoCAD offer various facilities and options to make the dimensioning of drawings easier and faster, as will be seen and guided throughout this chapter.

4.2 DIMENSIONING STYLES

Before inserting any dimensions in an AutoCAD drawing, several parameters such as type of arrows, lines, offsets, text height, etc. need to be set, following which the dimensions that are set adopt the set parameters. In the same manner, one drawing can consist of different parameters for the same attributes. This can be achieved by setting different dimension styles within the same drawing and then choosing which dimensions should fall under a specific dimension style. The dimension variables or parameters can be set by invoking the DIMSTYLE command which brings up the dialogue box shown on the left of Figure 4.1. From this dialogue box, a new style can be added using the **New** button and the current or default style can be changed using the **Modify** button to give the second dialogue box on the right of Figure 4.1. The discussion in this section will centre on this dialogue box

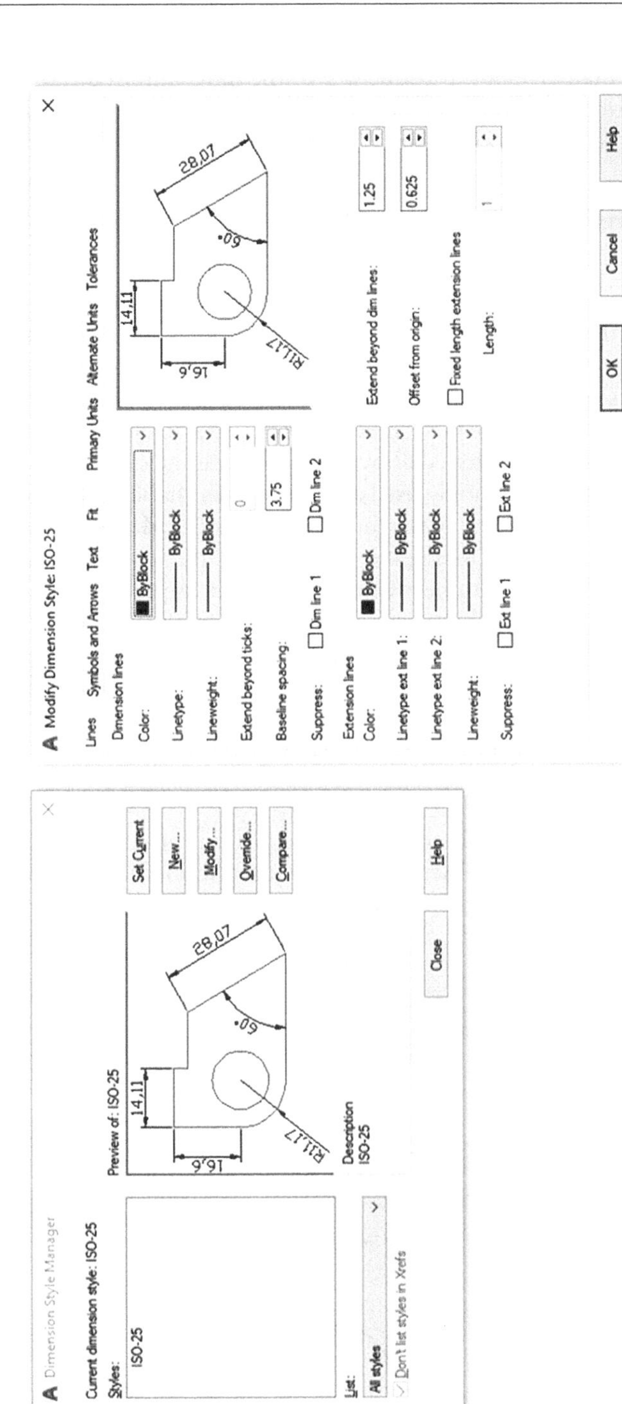

Figure 4.1 Dimension style manager dialogue box.

for setting the key parameters shown at the top of this dialogue box. This is however not exhaustive but introductory to the most important, the rest of which users can explore and change while viewing the adjustments on the preview of the second dialogue box.

4.2.1 Dimensioning attributes/terminology

Depending on which attribute the user wishes to set, modify or define, there must be a clear understanding of what they are setting, thus the importance of understanding the terminology used for the dimension variables and parameters. These can all be set using any of the radio buttons at the top of the **Modify Dimensions** dialogue box in Figure 4.1, listed as Lines, Symbols and Arrows, Text, Fit, Primary Units, Alternate Units and Tolerances. Figure 4.2 shows a portion of a dimensioned drawing to illustrate and explain the various meanings of the most commonly used attributes, thus enabling the user to set these appropriately under any of the radio buttons under **Modify Dimensions**.

What has been availed in Figure 4.2 are also the key and important parameters that users need to be aware of. However, as per Figure 4.1, there are various other peripheral parameters, ranging from lines used in dimensioning through arrows, text, fit, units and tolerances. Some of these will be used as the chapter progresses but the rest can be explored as these are alternatives.

4.2.2 Types of arrows

There are several types of arrows embedded within the AutoCAD system, the most common being the Arrow which is used in mechanical and other forms of drafting. The tick, commonly used in architectural drawings and the dot in electronic drawings, as well as several others, are available under

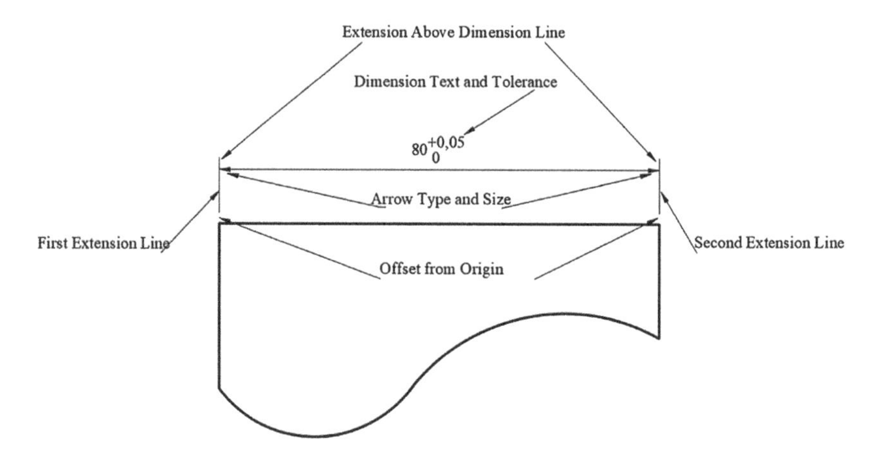

Figure 4.2 Dimension attributes and terminology.

the Symbols and Arrows radio button. Apart from setting the type, the user can also choose to have one type of arrow at the beginning of the dimension to a different one at the end, as well as setting the scale or size of the arrow. As these modifications of parameters are being executed under the Modify Dimension dialogue box, there is a preview window to enable the user to see whether the changes suit their requirements, on the failure of which further adjustments are made. Sometimes the preview window does not quite provide a satisfactory preview of how the dimensions will appear in the actual drawing. As such, users can accept the style defined and previewed and then proceed straight to the drawing to see if the final outlay is what is desired. This will also be dependent on the size of output drawing or paper but visibility is of utmost importance.

4.2.3 Text position and fit

There are several ways in which AutoCAD allows users to place their dimension text. The type of dimension in Figure 4.2 (placed above dimension) line is the standard adopted for the rest of the book in line with drawing conventions used in the book. The rest of the dimensions in the book are set in a similar manner such that the dimensions can be read upright or from the right, but always placed above the dimension line. There are other options available such as opening a gap on the dimension line and fitting the text therein. If the text height is not set prior to the drawing session, the text height can be set or even changed by varying the text height under the **Modify Dimensions** dialogue box.

4.2.4 Tolerances

From time to time, detailers are required to specify the allowances between which dimensions are allowed to fall. This is particularly useful for machine drawings where parts fit into each other, such as shafts and holes. AutoCAD offers the options for:

Symmetrical tolerance: Adds a ± expression of tolerance in which a single value of variation is applied to the dimension measurement.
Deviation: Adds a ± tolerance expression. Different plus and minus values of variation are applied to the dimension measurement.
Limits: Creates a limit dimension. A maximum and a minimum value are displayed, one over the other. The maximum value is the dimension value plus the value entered in the upper value.
Basic: Creates a basic dimension, which displays a box around the full extent of the dimension.

Depending on any of these settings, the appropriate tolerance can be set, over which the machinists have room to manoeuvre during production. These deviations are also dependent on what type of fit is required in the case of say a shaft fitting into a hole, the three being clearance (loose fit), transition (diameter of shaft and hole equivalent) or interference (tight fit). The rest of the dimension variables can also be adjusted to suit the user's requirements.

4.3 DIMENSION TYPES

While numerous dimensions of different types can be added to a drawing, there are basically seven types of dimensions at the user's disposal and the several variations within them are the ones that result in different layouts and orientations of these dimensions. Figure 4.3 shows an illustration of the seven dimension types and how they are positioned on a typical drawing, in relation to the terminology used in Figure 4.2. When a wrong type is picked, for instance when the user intends to dimension horizontally but accidentally picks points that are on a vertical line, this will result in a zero or null dimension, hence the need to take extra care in selecting the most appropriate from the seven. The seven types of dimensions, with the first three letters capitalized, as these are the letters used in conjunction with the **Dimension** command in the next section, are as follows:

1. ***Horizontal***: Restricted to any horizontal edges on a drawing (HOR)
2. ***Vertical***: Restricted to any vertical edge on a drawing (VER)
3. ***Aligned***: Drawn at any angle as long as it is aligned to a particular edge (ALI)

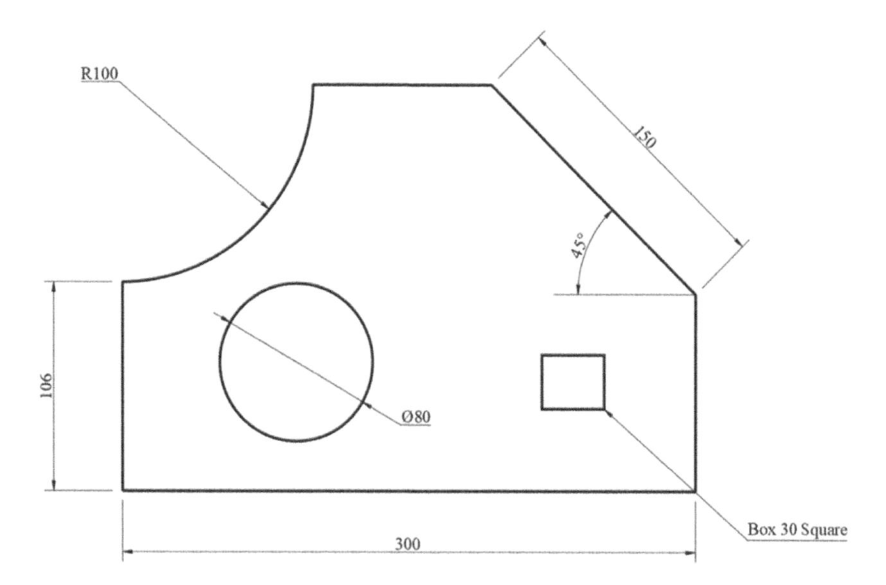

Figure 4.3 Illustration of the seven types of dimensions in AutoCAD.

4. ***Diametral***: Used for dimensioning diameters of circles (DIA)
5. ***Radial***: Used for dimensioning arcs by inserting radius of the arc (RAD)
6. ***Angular***: For dimensioning angles (ANG)
7. ***Leader***: For labelling any components within a drawing

4.4 INSERTING DIMENSIONS

Having set the variables as required under the Modify Dimension dialogue and satisfied that the preview shows the display of how the dimensions should appear in a drawing, the command, DIM can be used followed by the first three letters of the type of dimension required as shown in Section 4.3. For the horizontal, vertical and aligned dimensions, the start and end points of the dimension are selected followed by positioning it appropriately using LMB.

For the diametral and radial dimensions, the circle or arc must be selected at an appropriate position and immediately after the annotation for the measurement automatically appears, and this can be moved and inserted appropriately by selecting the LMB. The ISO standards are usually adopted for such dimensions as shown in Figure 4.3 and in particular for the diametral dimension to appear as a full dimension line spanning across the circle, the option, **Draw line between extension lines** under **Fit** and **Fine Tuning,** needs to be ticked on. For the angular dimension, a construction line may be necessary such as that shown in Figure 4.3.

The two lines intersecting can be selected anywhere along their length, one after the other followed by selecting an appropriate position to insert dimension. The leader dimension is really not a measurement dimension but a label for explaining certain components that may be in the drawing. In the earlier versions of AutoCAD, the leader dimension was inserted using the same procedure as the other six dimension types by the command DIM followed by LEA. However, the newer versions of AutoCAD such as AutoCAD 2021, simply require users to type in LEA followed by start point of the arrow pointing at objects within the drawing, then the position of the label, followed by the required annotation as shown in Figure 4.3.

4.5 MODIFYING DIMENSIONS

Apart from previewing how dimensions would appear after setting the variable parameters, users can also proceed to accept the setting and experiment by inserting dimensions in an actual drawing to see whether they would appear as desired. If they are not satisfactory, the Modify Dimension dialogue box can be recalled and adjustments can be made for the various parameters as described above until the dimensions appear as desired.

In addition, it should also be noted as explained earlier, that the two or more dimension styles containing different attributes can be saved in the same drawing and the user can specify which dimension style to use for any given set of dimensions.

4.6 APPLICATIONS: COMMON MISTAKES AND REMEDIES

Whether it is in manual or computer-assisted drawings, dimensions must conform to set standards and conventions, for uniformity and clarity to the users. However, users tend to mix these standards, leading to confusion. One of the good things about CAD and in particular AutoCAD is that it offers the user many options to choose from, such that standards from all parts of the world are available. With that in mind, users need to be familiar with their drawing standards and conventions, which they must religiously stick to, to avoid providing detailed drawings that end up confusing technicians on construction sites or production factories. Some of the common mistakes made by CAD users are discussed below together with suggested remedies for such mistakes.

4.6.1 Superfluous dimensions

There is a tendency among detailers and designers to provide many details, particularly dimensions, specifications and labels on engineering drawings. While this is good for clarity, it tends to cloud and make drawings complex and difficult to read and interpret. Engineering and technical drawings filled with too much information lose sight of true intent. Any component holds an unlimited number of dimensions for any part. Only a few of those dimensions affect how the part functions. Users must focus their detailed engineering drawings on the most important and impactful dimensions. If a section of a drawing is already dimensioned, say at the bottom, there is no need to re-dimension it at the top, otherwise it simply clutters the drawing. This is generally referred to as superfluous or unnecessary dimensions which users must try as much as possible to avoid. Providing such superfluous dimensions is also referred to as over-dimensioning and it may lead to confusion by those on site or in production by diverting attention to non-essential features, but instead focusing on those that are critical for the proper functioning of the component or machine.

Tolerances for parts that intersect and work together such as shafts and holes, can be established after selecting the base control and functional features of the machine under development. The variations that the part or component can tolerate can be established from expert analysis from designers such that sufficient but not unnecessary details for tolerances are provided. Such datum can be used for reproducible results.

4.6.2 Insufficient dimensions

In a bid to provide just enough dimensions to avoid superfluous ones described above, sometimes users also fall into the trap of under-dimensioning and providing less than the required dimensions. Naturally, this leads to delays in production as technicians are forced to communicate with detailers back and forth to establish the missing dimensions. There is also a high risk and potential for technicians to estimate missing dimensions, thus compromising on the final product and its functions, let alone redoing the production, which can be costly to the company.

It is understandably time-consuming to ensure that sufficient dimensions are provided while at the same time avoiding clutters caused by too many unnecessary dimensions. In circumstances such as this where a user is faced with the challenge of superfluous or insufficient dimensions, they would rather keep superfluous dimensions as long as the drawing is not too detailed and complex. After completing a drawing and when detailing starts, it is advisable to look at the various edges and parts to ensure that they are adequately dimensioned. In the same process, dimensions that are duplicated or repeated can be removed. At the end of it all, it is the skill to balance between superfluous and insufficient dimensions that users need to have and in so doing, the focus should be on the most important dimensions to do with the function of a component or machine, such as specifying tolerances within which production or site personnel can comfortably work with.

4.6.3 Crisscrossing dimensions

Virtually all drawing conventions require that dimensions are inserted outside the drawing or on the periphery of the drawing. In the same vein, crisscrossing dimensions should also be avoided as they create confusion. However, in some instances internal dimensions cannot be avoided depending on how complex a drawing is. Generally, it is advisable to avoid such dimensions crossing each other. One way in which this can be avoided is to dimension outwards by starting with the smaller dimensions close to the component or machine and then moving outwards to end with the overall dimensions. It is also advisable to leave a sufficient gap between the start of the extension lines and the component (offset from the origin as shown in Figure 4.3), otherwise without the gap, the dimension and extension lines may confuse the reader in terms of distinction.

4.6.4 Incomplete specifications

The advent of CAD was meant to improve productivity in the development of engineering drawings. This book focuses on the development of skills for organizations to realize the benefits of investing in CAD. Principally, all the tools discussed in the book are based on developing shortcuts but still efficiently producing drawings. However, while some shortcuts may save time, some may actually introduce new challenges. For instance, generalizing tolerances may lead to unwarranted failures of machine components, because one size does not necessarily fit all.

Another shortcut that often leads to problems is the use of coordinate dimensioning, where a series of 'chained' dimensions are created having a common origin and single dimension line. It is advisable to avoid such baseline dimensions, as they also tend to create clutter and difficulties in interpretation as this does not emphasize form or orientation control of the feature. This is one aspect in which incomplete specification of the feature results from baseline dimensioning (Barari, 2009). Geometric Dimensioning and Tolerancing (GD&T) can be used to control form and orientation (Stites & Drake, 1999). These controls include multiple elements of functionality within one callout, thus reducing dimensioning, specifying functionality and alleviating overly tight tolerances.

4.6.5 Over-constrained tolerances

Over-constrained tolerances in technical drawings can also create confusion for technicians in production or on site, thus escalating costs. Sometimes due to lack of foresight or in some instances just wanting to ensure safety and protecting themselves, engineers make tolerances over-tight. As alluded to above, coordinate dimensioning has its limits but this can be resolved by defining location control with a true position. True position implements a diametric tolerance zone around a feature and the true position evaluation opens up the boundary and does not confine dimensions to a square tolerance. As mentioned for incomplete specifications, GD&T true position defines a diametric tolerance zone of movement that can allow near 50% more variation in the hole position than with coordinate dimensioning. The allowable tolerance can be increased while focusing on how the components fit into each other. Such true positions in technical drawings give personnel in production more room to manoeuvre.

4.7 SUMMARY

Well-defined and detailed engineering and technical drawings are an absolute necessity and should translate functional requirements into the intended design, whether it is for the production factory, construction site or power generation plant. This chapter focused on the different types of dimensions used in AutoCAD, how to insert and modify them and ultimately ends with applications and what users need to watch out for to avoid superfluous and insufficient dimensions. GD&T helps to unify the purpose of engineering drawings and the function of the end product. In addition, GD&T also helps to overcome challenges brought on by coordinate dimensioning, thus creating clear representations of how features of components are integrated into the product assembly.

GD&T enables simulation of functionality through measurement and conveying functional design purposes through geometric feature control rather than coordinating the size and location of a feature, thus achieving repeatability and reproducibility in detailed engineering drawings.

4.8 REVIEW EXERCISES

1. Outline, explain and illustrate the seven types of dimensions used in CAD, paying particular attention to the variable parameters that must be defined before any of the dimensions are inserted in drawings.
2. Using an illustration, explain at least five dimension attributes that should be set and defined under the Modify Dimensions dialogue box to ensure that dimensions and drawings appear as desired.
3. Outline at least five common mistakes that CAD users make in dimensioning in real practice and applications. For each, provide possible remedies as implemented in AutoCAD.
4. Figure 4.4 shows an incorrectly dimensioned engineering drawing. List at least five errors from this drawing, explaining each and also outlining how to resolve them.
5. Reproduce Figure 4.4 in AutoCAD and provide sufficient but not superfluous dimensions that are properly inserted in line with drawing conventions.

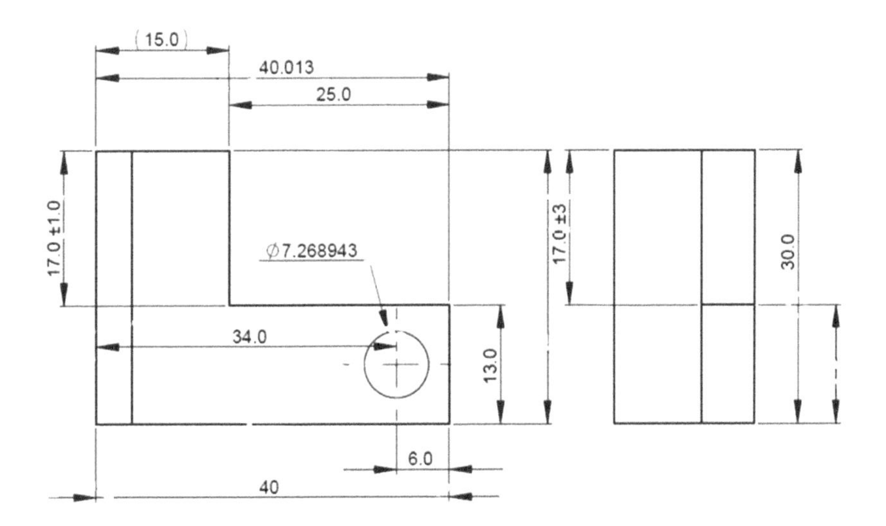

Figure 4.4 Incorrectly dimensioned engineering drawing.

Chapter 5

Layers

5.1 INTRODUCTION

Technical and detailed engineering drawings are unavoidably very complex and cluttered with detailed information regardless of whether they are mechanical, production, architectural or any other form of drawing. The multitude of information in detailed drawings ranges from the outline of the drawing itself, hidden details, centre lines, dimensions, specifications, labels, etc. Although visualization of the drawing under development can be improved by zooming in and out of the drawing, this has limitations due to the sizes of CAD screens commonly in use. This makes both visibility and analysis difficult as one navigates through the drawing. In some cases, artisans on site or in the production factory also tend to use their mobile phones to view the drawings as it may be an inconvenience to bring a laptop or bigger CAD screen on site, hence the need for ways in which such complexities can be addressed. Invariably, this entails either printing a portion of the drawing or maintaining the entire drawing but leaving out certain other details in order to reduce the complexity. However, this compromises the end product. In a bid to make this workable with CAD systems, some organizations actually go to the extent of investing in larger screens such as 50 inches and above as opposed to the regular 15 or 17 inch screens.

Most CAD packages nowadays, including AutoCAD, provide the users with the flexibility of splitting the drawings into different levels (layers). To understand this better before embarking on the whole concept of CAD layers, these levels can be visualized from the example of architectural plans of multi-storey buildings where one floor lies directly on top of another. However, architectural drawings of this nature cannot obviously be superimposed on each other as they are in real life. Instead, each level or floor is handled individually to avoid a clutter of details.

The CAD facility for layers not only allows the user to split the drawing so that one level can be handled individually but also offers options to manage other drawings such as machines or components in such a way that common details can be grouped together and analysed separately from the

DOI: 10.1201/9781003288626-5

other details. Such options include the ability to temporarily turn some layers on and off, lock or unlock them to avoid accidental changes, as well as freezing or thawing. The sections that follow in this chapter describe these tools in detail in order to enhance the skills in using CAD as well as to justify investment in these systems for a good return on investment.

5.2 LAYER PROPERTIES MANAGER

Before starting any drawing in CAD and depending on how detailed and complex it is, it is advisable to establish some layers over which details of the drawing can be inserted and managed. The Layer Properties Manager (in newer versions) or previously, the Layer Control Dialogue box, can be loaded by using the Layer (LA) command or selecting the Layer icon (stack of three papers) on the slide menu to bring up what is displayed on the snapshot in Figure 5.1. Users can define and add as many layers as they wish but need to be careful that the more layers there, the more confusing the drawing can become. The important and critical areas of this layer property manager will now be described in detail in the following sections.

At the top of the Layer Properties Manager, are four stacks of paper:

1. Creates a new layer (yellow star),
2. Creates a new layer that automatically freezes all layout viewports (blue star),

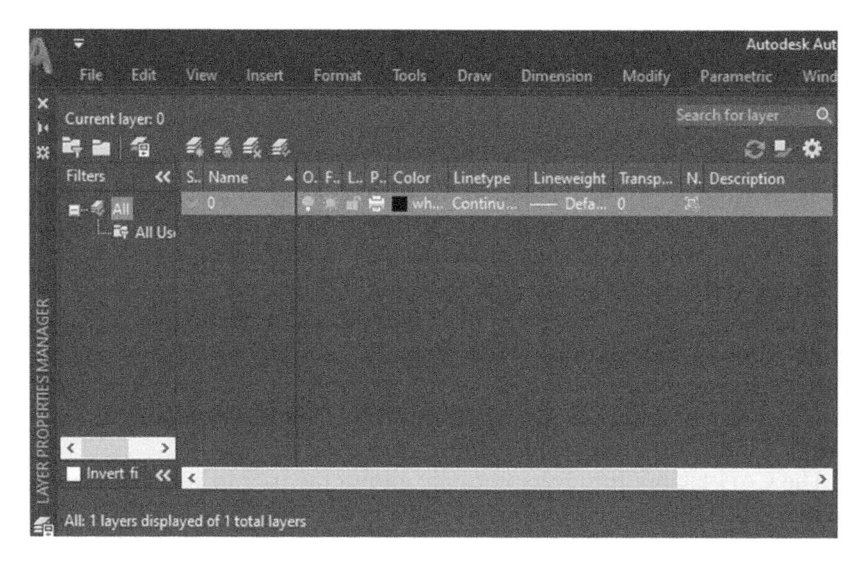

Figure 5.1 Layer properties manager.

3. Deletes a selected layer but this can only be done on an unreferenced layer (red cross)
4. Sets selected layer to be the current layer (green tick), over which additions will fall.

Every time a new layer is created, it is added to the list and allocated an automatic name, Layer1, Layer2 etc. The user can highlight the name and write over to insert their preferred name of the layer. Alongside the list of layers will appear the status for each:

Turned ON or OFF (bulb lit or off),
Frozen or Thawed (sun ON or OFF),
Locked or Unlocked (padlock locked or unlocked),
Plottable on not (printer ON or greyed),
Colour (selected colour for the particular layer),
Linetype (selected linetype for the particular layer) and
Lineweight (selected lineweight for the particular layer).

Once all the layers have been created, the Layer Properties Manager can be turned off and the development of the drawing progresses. During the process and if required to move from one layer to another, the Layer Slide Menu can be used but if certain properties need to be adjusted then the Layer Properties Manager can be recalled and appropriate adjustments made before proceeding further.

5.2.1 Turning layers ON or OFF

Any layer created in a drawing can be turned ON or OFF as long as it is not the current layer. Layers that are turned ON are visible on the screen and can be modified but layers that are turned OFF are temporarily invisible and thus cannot be modified or tampered with. This facility is meant to avoid accidental changes to details on a layer that is turned OFF. Because such layers are invisible, they can also not be used as a reference to any changes that might be taking place in a drawing.

5.2.2 Locking or unlocking layers

Similar to turning layers ON and OFF, except for the current layer, any layer created in a drawing can be locked or unlocked. Layers that are locked are visible on the screen but no changes such as erasing, trimming, extending, etc. can be carried on such layers. They can however be used as a reference to changes or developments on other layers. Locked layers need to be unlocked first before executing any changes to them. This facility is particularly useful when one is working on other layers but wishes to use a locked layer as a reference point for such tools as object snap modes like endpoint, midpoint, centre, intersection, etc. This is therefore the primary difference between turning layers ON & OFF and locking & unlocking layers.

5.2.3 Freezing and thawing layers

Likewise, for turning layers ON and OFF or locking and unlocking them, layers can also be isolated from any developments that may be taking place in a drawing. When working with drawings with lots of layers, freezing unnecessary layers can speed up display and regeneration. Entities on a frozen layer are not considered during Zoom Extents. However, frozen and thawed layers are both visible and can be used as reference points. To unfreeze layers or thaw them, the entire drawing will then be automatically regenerated.

5.3 PRACTICAL APPLICATIONS

Virtually all disciplines that use drawings, whether manual or computer-assisted, benefit from the use of layers. This is due to the complexities and amount of details contained in typical engineering drawings. In architecture, particularly multi-storey buildings, several floors can be developed in one drawing but each floor is separated and isolated into respective layers. When working on floor 3 for instance, floors 1, 2 and 4 can be turned off or if needed for reference purposes, they can be locked but still visible. Since such floors will be superimposed on each other, the different levels can be displayed separately on the graphics window and better still distinguished from each other by using different colours.

For mechanical and electronic drawings that contain so many minute components and details, these can be grouped and isolated in layers that can be manipulated to the desired layout. For instance, for a machine that consists of a housing and several components inside, fasteners such as bolts and nuts, screws etc. can be grouped together in one layer, dimensions in another, specifications in yet another, and so on. For electronic drawings such as Printed Circuit Boards, depending on the quantities for each of the components, resistors can be on their own layer, diodes on another, connectors and switches on yet another layer, and so on.

Alternatively, depending on what function the machine or PCB is supposed to perform, components can be grouped on a common layer to allow for necessary technical modifications that may be required. Layers are thus, the primary method for organizing the various entities in a drawing by function or purpose. Layers can improve the visibility of details on a complex drawing and thus reduce the chances of compromising the output by errors that may arise from a mix-up of specifications or dimensions by hiding information that may be unnecessary at a given point of development, be it in the drawing office, site or production plant.

Quite often, complex and detailed drawings can comprise numerous layers. A layer filter can be used to limit the display of layer names in the Layer Properties Manager and in the Layer control on the Slide Menu. Layer property filters can be created based on the attributes mentioned earlier such as different name, colour, linetype, etc. The naming of layers in a certain manner can help to group layers with similarities. Layer group filters can also be created if specific layers are chosen to include in the group. Such groups could be fasteners in mechanical design, electronic components in electronic circuit design or plumbing materials in architecture.

Much as it is desirable to reduce complexity in drawings by improving visibility, there are possible chances that invisible layers (turned OFF) may compromise discussion of what may be visible and thus picturing an incomplete drawing. In addition, once a layer is used as a reference to another, it will not be possible to modify the reference layer, such as deleting it.

5.4 SUMMARY

Although most drawings used in this book are simple and straightforward, mainly for the purposes of users to grasp the techniques and skills required to use CAD beneficially, real-world and detailed engineering drawings can be quite complex, more so when viewed from the common 15–17 inch screens that are used for CAD. Drawings in CAD can be split into levels or layers in order to reduce their complexity and improve visibility. This chapter focused on how to create and use layers in AutoCAD in order to reduce the visual complexity of detailed drawings and to improve display performance by hiding information that may not be necessary. This is accomplished by turning layers ON or OFF, locking or unlocking as well as freezing and thawing, thereby easing the management and organization of drawings for use in drawing offices, production plants or construction sites. These tools also enhance and justify the need to invest in CAD systems.

5.5 REVIEW EXERCISES

1. Explain at least five reasons why layers are used in CAD, giving practical examples.
2. What are the main properties that can be attached to a layer in a drawing under development?
3. Users have the flexibility of turning layers ON and OFF, locking or unlocking them or freezing and thawing layers:
   i. Explain what each pair of tools of the three does.

   ii. In terms of visibility, freezing and thawing layers are the same as locking and unlocking layers. Explain the subtle differences between the two pairs, giving examples where both may be used.

4. Several disciplines in engineering can benefit from investing in CAD. Focusing on the use of layers, list five such disciplines and explain how layers can be used to reduce the complexity of drawings and improve visibility by listing typical layers that can be created for each discipline and what components to add to those layers.
5. List and explain three limitations or disadvantages of using layers in CAD and in addition, explain how such limitations can be overcome.

Chapter 6

Orthographic and isometric projections

6.1 INTRODUCTION

There are many ways in which objects in real life can be modelled and represented. Modelling is a word like design, which is used rather loosely, in everyday language, ranging from components or parts of machines to the construction of scale models that represent real objects. Modelling can be a much more formal activity when a scientist, engineer or architect wishes to explore the nature of some physical process or understand how to construct or operate some physical object or system (Norman et al., 1990). A model is thus a substitute or representation of the real object, hence the importance of doing it properly in order to maintain its real-life properties. There are different types of models that can be used to represent objects. Each type has its own characteristics, which are useful to the designer, for example, manufacturability, aesthetic appeal and shape in general. However, conventionally the majority of detailed engineering and technical drawings are orthographic and to a lesser extent isometric (pictorial). The use of these is dependent on what is to be achieved. Certain principles and conventions need to be adhered to in order to maintain the same interpretation in the drawing office and on-site or in production.

Orthographic and isometric projections are the two most commonly used projections in engineering and technical drawings. This is regardless of whether the drawings are developed manually or with the assistance of a computer. This chapter focuses on developing user skills for quickly generating such types of drawings using AutoCAD and based on engineering drawing conventions in order to further justify the investment in CAD systems. However, it should be noted that although isometric drawings are pictorial in nature and appear to represent objects as visualized in real life, they are not 3D objects. 3D modelling will be handled from Chapter 8 in which AutoCAD has further enhanced the application of their software by introducing tools to automatically generate orthographic projections from 3D models. This chapter will thus be limited to the pictorial representation of objects and their associated orthographic views.

DOI: 10.1201/9781003288626-6

The orthographic projection system is used to represent real-life objects in 2D planes. The orthographic projection system utilizes parallel lines to project pictorial views onto 2D planes. In a way, this is how the word *Orthographic* was derived from the two words, *Ortho* meaning orthogonal or at right angles and *Graphic* meaning picture, hence images or pictures at right angles to the viewing plane. It is assumed that users of this book will be familiar with such conventions, derived from manual drawings. The display of orthographic projections is governed by the type of angle of projection selected as discussed in the next section.

Isometric projections on the other hand are pictorial representations of objects but strictly are not three-dimensional as the pictorial views are displayed on a 2D plane. The word *Isometric* is derived from the two words, *Iso* meaning same and *Metric* meaning measurement. Isometric views are thus pictorial views drawn at 30° to the horizontal with the same measurement close to the viewer as it is further away from the viewer, as opposed to Art where measurements close to the viewer tend to be visually larger than those further from the viewer.

6.2 ANGLES OF PROJECTION

The orthographic views that are derived from an object or its pictorial at right angles to the viewing plane are not simply placed at random but their positions are dependent on which type of projection system would have been chosen, whether first or third angle projection. Engineering drawing is a means of communication and universal language that must be understood by all involved in the practice of engineering from design engineers through technologists to technicians.

6.2.1 First angle projection

The convention for positioning orthographic views derived from viewing an object at right angles to it, and placing the elevation to the opposite side from where the object is viewed is referred to as first angle projection, and thus appears in the first quadrant as shown in Figure 6.1. The object is positioned at the front of a vertical plane and top of the horizontal plane. The object is placed between the observer and projection planes.

6.2.2 Third angle projection

In the third angle projection, the object is placed in the third quadrant and behind the vertical planes and bottom of the horizontal plane as shown in Figure 6.1. The resultant elevation or view obtained is placed on the same side from where the object is viewed from. The projection planes come

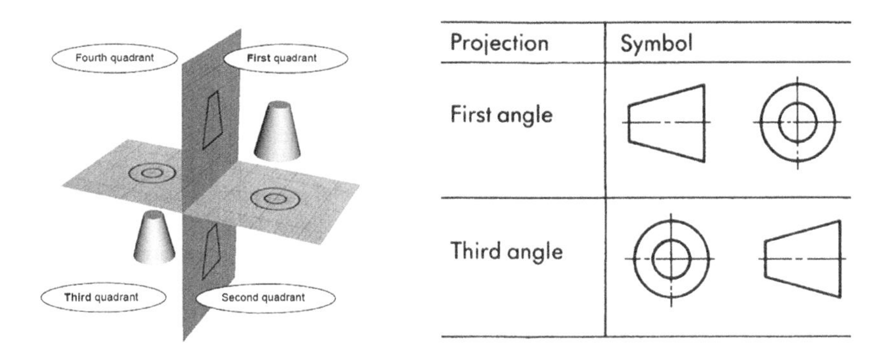

Figure 6.1 First and third angle projections.

between the object and the observer. The plane of projection is taken as transparent in third angle projection.

Apart from showing the quadrants, Figure 6.1 also shows the symbols adopted for the two types of orthographic projections. Where such views are displayed adjacent to each other, it is a conventional requirement to insert the appropriate symbol to clearly indicate which type of projection was used in the particular drawing.

6.2.3 Differences and applications of projections

First angle projection is widely used throughout the world but mostly in Europe, hence it is sometimes referred to as the European projection. Third angle projection, upon which all exercises and examples in this book are based, is the system commonly used in North America and hence, sometimes referred to as the American projection (Simmons et al., 2020). In the rest of the world, particularly in the British Isles and former British colonies, both systems of projection are regularly used.

Generally, students of engineering and technical drawing are trained to be conversant with both systems of projection in order for them to be versatile for future employment around the world. In fact, in the current British and ISO Standards as well as the standards for most countries in Southern Africa, both systems of projection are acceptable. However, these should never be mixed, otherwise, drawings from the design office to the construction site or production plant can easily be misinterpreted, hence the importance of always appending the appropriate symbol on all drawings.

Table 6.1 lists a comparative analysis of the first and third angle projections, with reference to the quadrants shown in Figure 6.1. Beginners or learners of technical drawings often get confused about these systems of orthographic drawings. However, it should be noted that first and third angle projections are nothing but ways of describing what an object looks

Table 6.1 Comparison of first and third angle projections

	First angle projection	*Third angle projection*
1	Object is positioned in the first quadrant.	Object is positioned in the third quadrant.
2	Object is positioned between the plane of projection and observer.	The plane of projection is placed between the object and the observer.
3	Plane of projection is opaque.	Plane of projection is transparent.
4	Front view is at the top of the horizontal axis.	Front view at the bottom of the horizontal axis.
5	Top view at the bottom of the horizontal axis.	Top view at the top of horizontal axis.
6	Right view is on the left side of vertical axis.	Right view is on the right side of vertical axis.
7	Left view is on the right side of vertical axis.	Left view is on the left side of vertical axis.
8	Widely used in Europe, India, Canada, the UK and its former colonies.	Widely used in the United States, Australia, the UK and its former colonies.

like from different directions and orientations. Both systems are used for multi-view projection of 3D objects using a series of 2D drawings.

The principal planes of the object are used to project different views of the same object from different points of visualization. Overall, six different sides can be drawn consisting of six orthographic views commonly referred to as the principal views.

Both methods of orthographic projection result in the same six principal views of the object except for the arrangement of views and the state of the plane of projection. In the first angle projection, the plane of projection is believed to be opaque or non-transparent. The object is placed in front of the planes and each view is pushed through the object which places the vertical plane behind the object and pushes the horizontal plane underneath.

In the third angle projection method, the plane of projection is transparent and the object is placed below the horizontal plane and behind the vertical plane (Simmons et al., 2020). Users of CAD are strongly advised to understand bot systems of projection in order to be able to read or use drawings generated from elsewhere. It is therefore imperative to ensure that one appends the appropriate symbol to every drawing and also look out for this symbol first before using any drawing generated from elsewhere. This helps in comprehending what would have been communicated from source, either by the design engineers or detailers, otherwise without doing so, may result in a completely different object being analysed or developed. If users are more familiar with a particular projection, it is probably best to quickly reposition the given orthographic projections in such a way that there is a full appreciation to avoid any mix-ups as the development or production progresses.

6.3 ORTHOGRAPHIC PROJECTIONS FROM PICTORIAL VIEWS

Generally, orthographic views are derived from viewing objects at right angles to any given plane. Ultimately the elevations that can be extracted from an object are; Front, Left and Right Side, Top (Plan) and Bottom. In common practice, at least two of these orthographic views are required to describe an object completely but more may be necessary depending on the complexity of the object being modelled. Detailed drawings (orthographic elevations) must be adequately dimensioned to ensure that the correct information is communicated from the design office to the production plant or construction site. In addition, while there should be just enough dimensions to completely describe the object, care must also be taken to ensure that there are no superfluous or unnecessary dimensions that will simply clutter the drawing and create confusion. Normally, most drawing conventions require that sectioned elevations do not show hidden details or lines otherwise such lines may also create confusion as they intersect with the hatch or section lines, thereby cluttering the drawing as well. Although users are advised to understand both systems of orthographic projection, the third angle projection system is adopted in this book as the standard.

Knowledge of first angle projection is useful to be able to interpret drawings created elsewhere or in some cases where the first angle of projection is adopted as the standard by an organization or country. To generate orthographic elevations from a given object, either in 3D or in isometric/pictorial view, the steps followed in manual drawings are similar. Firstly, it is important to ascertain the overall dimensions of the object, that is, length, width and height. It is also important and necessary to establish the orthographic elevations required. This will either be specifically stated but if it is not specified then the user should determine the most appropriate elevations that will describe the object best. If it is a simple object then the Front and Plan views may be adequate but if it is complex, an additional side view is added.

Having established what elevations are required, the overall rectangles, derived from the length, width and height of the object, for the respective views are then drawn, adequately spaced from each other. At this point, the user may want to use thicker polylines for the outlines in order to distinguish these from construction lines. It should also be noted that as the construction of elevation progresses, there will be lots of construction lines, but these should be erased as and when they are no longer necessary otherwise they also create clutter on the screen and this may lead to confusion in delays in developing the required orthographic elevations.

Within the constructed rectangles, details from each face of the object are then inserted into the appropriate rectangle, be it the view derived from the Front, Top or Side of the object. As in manual technical drawing, a 45°

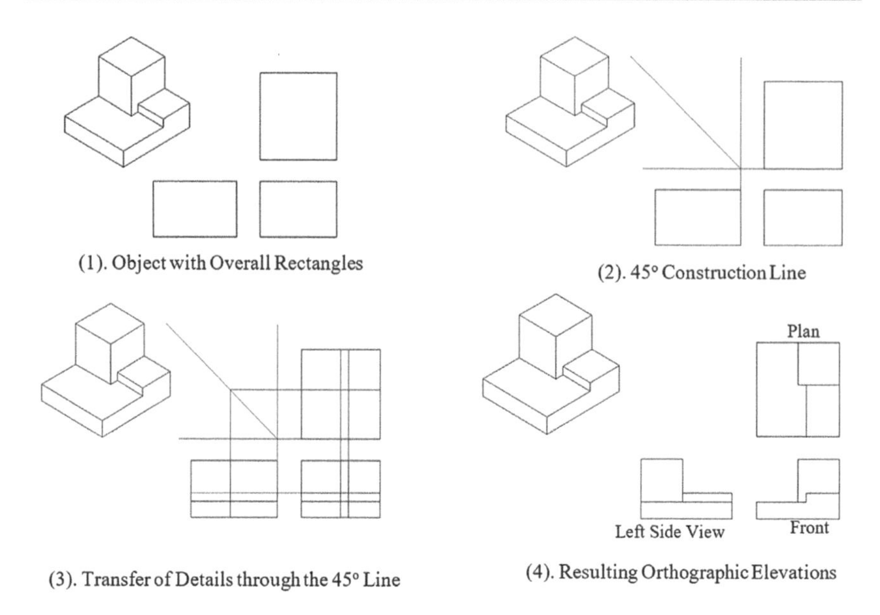

Figure 6.2 Isometric to orthographic elevation steps.

construction line can be used to transfer details from one elevation to the other as shown in Figure 6.2. Where necessary, trimming is then carried out and unnecessary lines, edges or construction lines are removed to remain with just the required elevations. Although this is the usual practice in industry, students may be required to leave construction lines visible in order for instructors to ensure that they used the correct procedure.

6.4 ISOMETRIC PROJECTIONS FROM ORTHOGRAPHIC VIEWS

Frequently, engineers need to illustrate drawings pictorially for clarity. Detailed orthographic drawings are preserved and can only be well understood by technical staff trained in the field; the only form of communication that other people can easily understand is by way of pictures or pictorial views, hence the use of isometric views. As alluded to earlier, although isometric views appear as 3D objects, they are actually models represented in 2D planes. For a company that manufactures and sells products, their brochures will be best presented in pictures, pictorial or isometric views for customers to have a good appreciation of what they have on offer. Isometric views refer to the pictorial representation of objects drawn at 30° to the horizontal as shown in Figure 6.3. As alluded to earlier, 'ISOMETRIC' refers to the same measurement of edges for those close to the viewer and those further away.

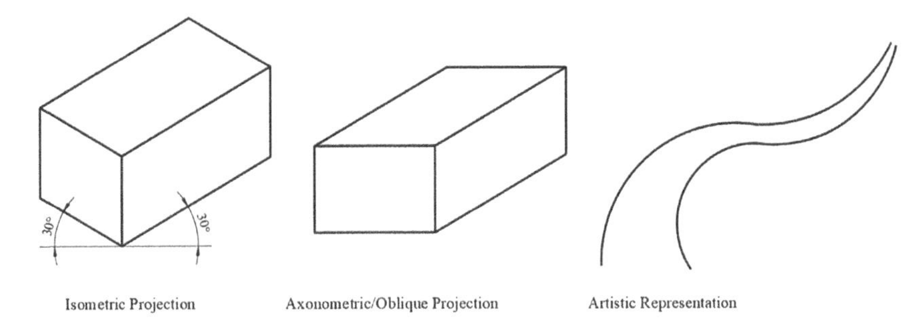

Figure 6.3 Pictorial representation of objects.

This is however different from how objects are viewed in Art, where the edges close to the viewer are close to the true lengths and those further away, are narrower as shown in Figure 6.3. Isometric views should also not be mistaken for axonometric or oblique projections. Axonometric projections are a type of orthographic projection used for creating pictorial views of an object, where the object is rotated around one or more of its axes to reveal multiple sides. Oblique projections are parallel projections in which the lines of sight are not perpendicular to the projection plane.

Commonly used oblique projections orient the projection plane to be perpendicular to a coordinate axis, while moving the lines of sight to intersect two additional sides of the object. Axonometric and oblique projections are not commonly used in engineering modelling and representation of objects but are a useful tool when the user requires some flexibility in the orientation of an object. This section is however limited to isometric projections as these are a fixed representation of objects and are commonly used in engineering drawing and design. In order to construct isometric views in AutoCAD, the cross-hair needs to be changed from the regular cross to that which will be oriented in any of the three isometric planes, that is, Isoplane Left, Isoplane Right or Isoplane Top.

The command Snap (SN) is used from where Style is selected and then Isometric. The snap distance can be set to any suitable distance or even left out (0.00). The cross-hair then changes its colour and orientation to one of the isometric planes. Drawing lines with ortho turned ON or pressing the shift key will produce lines that are parallel to the chosen isometric plane.

To toggle the cross-hair and change to another isometric plane, Ctrl-E is used and this can be done even in the middle of executing another command. The snap can also be turned ON/OFF using object handlers. The snap spacing is the distance between the snap or grid points. The aspect value allows users to set a different spacing for vertical and horizontal snap points, otherwise the default is the same value for vertical and horizontal points. To return to the regular cross-hair and display, the same procedure can be followed but this time, selecting Standard Style. The Rotation option allows the setting of snap points at an angle to the x-axis.

6.5 BOXING METHOD

In a similar fashion that was used to develop orthographic projections from a pictorial or isometric view, this section focuses on developing isometric views from given orthographic projections. This is commonly referred to in technical drawing as the boxing method. Firstly, a suitable isometric viewpoint that clearly depicts the maximum features of the object is determined. The user also needs to ensure that the correct isometric plane (Left, Right or Top) has been selected to construct the overall isometric box that will just fit the details provided, derived from the overall length, width and height of the object before constructing details onto the isometric planes. It must also be noted that circles in isometric planes appear as ellipses and as such, their construction should be done correctly by **Ellipse** followed by the **Isocircle** commands.

The other construction tools such as positioning the isocircle centred in a particular point can be followed. Object handlers such as ORTHO are useful for drawing lines parallel to the isometric planes only. It might be necessary to construct the isometric box and the respective edges using thicker polylines to distinguish these from any construction done with orthographic projections. The step-by-step procedure for constructing an isometric view from given orthographic views for a typical object is illustrated in Figure 6.4.

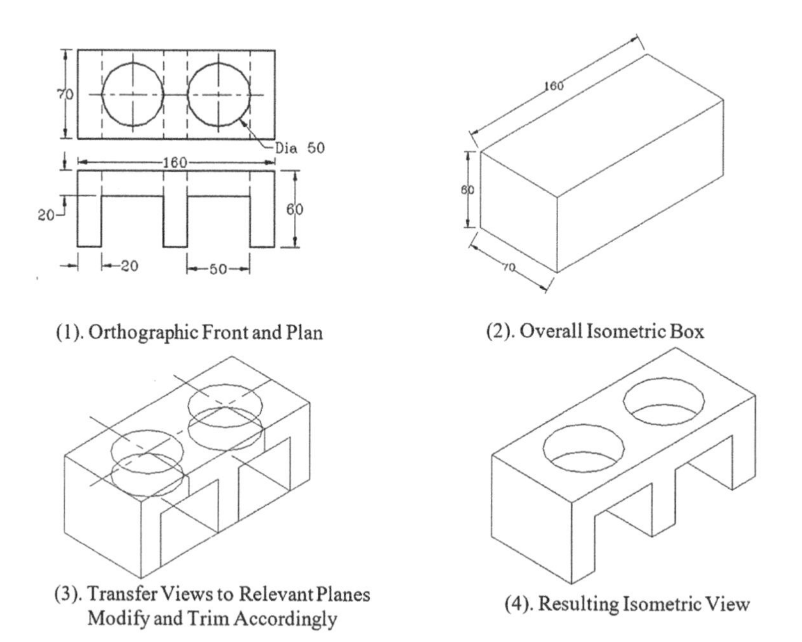

Figure 6.4 Boxing method for constructing isometric views.

6.6 APPLICATIONS AND LIMITATIONS

Isometric drawings are frequently used in technical drawing to represent objects that seemingly appear as 3D objects but in reality are modelled on 2D planes. This is one quick way in which objects can be modelled for illustration and understanding by non-technical personnel who ordinarily find it difficult to read, understand and interpret detailed orthographic projections.

Even for the technical engineering personnel, it is a quick way in which models can be developed and appreciated by just visualizing what is on display. Isometric drawings, commonly referred to as isometric projections, are a good way of showing measurements and how components fit together. Unless it is absolutely necessary, isometric views are meant to remain as simple as possible to ensure that designers, technicians and customers alike, understand them quickly without the need for dimensions that may clutter the presentation and create confusion. In some drawing conventions, isometric drawings should actually be left as pictorial views but with no dimensions.

Unlike perspective drawings, isometric projections don't get smaller as the lines go into the distance as shown in Figure 6.3. Axonometric or oblique projections are perspective views that can also be used to model and represent objects from any given orientation or perspective. They are more artistic and thus rarely used in technical drawings. While isometric or perspective projections are a good representation of objects, they are limited to just displaying the shapes that appear as 3D objects but on 2D planes. Therefore, they do not contain all the information about the object such as materials, mass moments of inertia, weight, surfaces etc., let alone the z-coordinate that would truly define a 3D object.

3D objects and modelling are handled from Chapter 8 onwards and as it will be seen, these types of models are not as easy and quick to create as 2D objects, apart from the amount of processing and space that they occupy. This is probably why some practitioners prefer the quick development of pictorial objects using isometric projections. Although 2D projections are less cumbersome to generate, a clear understanding of 3D modelling would be an added advantage for users as it will also be seen from Chapter 8, that 2D orthographic projections can automatically be derived from 3D objects, without the need to go through the processes covered in this chapter.

6.7 SUMMARY

The development of detailed 2D technical and engineering drawings makes use of two methods of projection – First Angle Projection and Third Angle Projection. This chapter covered, with clear examples, the rules for producing drawings for either of the two projection methods and symbols employed for the chosen method. The two methods of projection are in use in different parts of the world and thus advisable for users to be familiar with both methods of projection. Step-by-step procedures were also provided for generating orthographic views from given isometric or pictorial views as well as the use of the boxing method to create isometric views from given orthographic projections.

Although axonometric or oblique views are also useful representations of objects, their perspective nature makes them less useful for technical and detailed drawing presentations. Pictorial and isometric views are useful quick ways for communicating information on possible designs through brochures but are limited to only showing the shape without details such as materials, surfaces and weight, aspects that will be dealt with under 3D modelling in later chapters.

6.8 REVIEW EXERCISES

1. Using illustrations and quadrants, explain the difference between First and Third Angle projects, indicating why engineering practitioners need to be familiar with both systems of projection.
2. While some designers view isometric projections as three-dimensional, they are strictly not three-dimensional. Explain why this is so and provide practical examples where isometric projections are used in real practice.
3. Using illustrations, explain the source of the words Orthographic and Isometric.
4. Explain how circular objects are handled and inserted in isometric planes in a typical CAD system such as AutoCAD.
5. Figure 6.5 shows an isometric projection of an engineering casting. Study the figure carefully and then answer the following questions in AutoCAD:

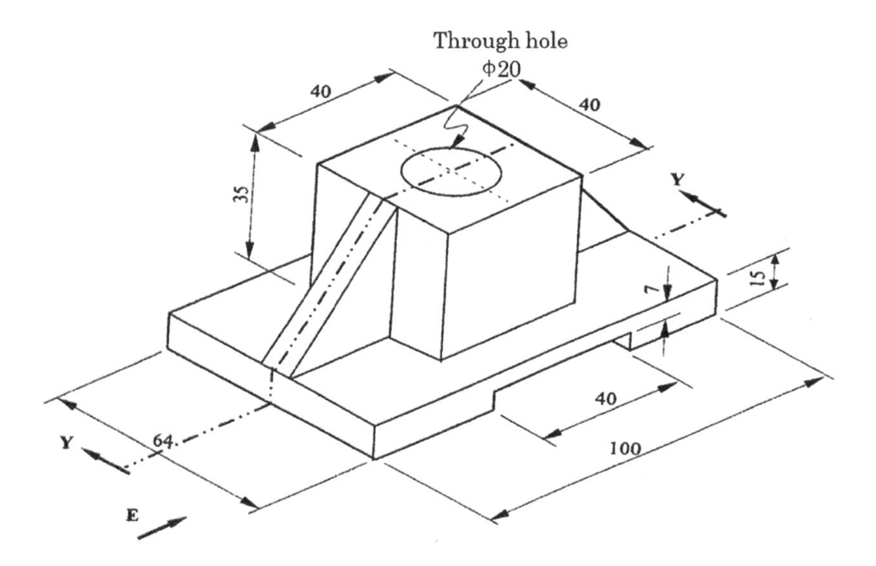

Figure 6.5 Isometric projection of an engineering casting.

Table 6.2 Layers for the engineering casting

Layer	*Linetype*	*Colour*
Section	Continuous	Yellow
Dimensions	Continuous	Red
Centre	Centre	Blue
Hidden	Dashed	Green
Text	Continuous	Magenta

a. Start AutoCAD, then create the following layers as shown in Table 6.2.
b. Using appropriate layers from Table 6.2, produce the following orthographic projections, clearly labelled and positioned using Third Angle Projection:
   i. Sectional front elevation taken on the cutting plane Y-Y
   ii. Left side elevation in the direction E
   iii. Plan
c. Incorporate the minimum number of dimensions on your projections to fully describe the casting. Any dimensions not given in Figure 6.5 can be estimated relative to those that are given.

Chapter 7

Blocks and attributes

7.1 INTRODUCTION

Other than lines, circles, text etc., most engineering and technical drawings consist of standard symbols that are used to represent standard components in everyday use. These range from fasteners in mechanical design, doors and windows in architecture, valves and gates in hydraulics, diodes, wires and resistors in electronic circuit designs. By and large, these symbols are basic structures with just lines and arcs and are repeatedly used from one drawing to another. Reproducing these repeatedly from one drawing to another can be time-consuming and costly. Developing a facility to create these once and store them in appropriate databases where they can be retrieved and used in any other drawing is one of the many facilities available in CAD systems such as AutoCAD. This removes the time-consuming and costly process of recreating them every time they are required in other drawings. AutoCAD refers to these symbols as blocks and this chapter explains how these blocks can be created together with their attributes (properties) and then stored for future use in other drawings. AutoCAD allows users to create such symbols derived from the standard symbols available in use in the multitude of engineering disciplines.

The information stored as properties or attributes of the blocks such as the rating of capacitor in electronic circuit design, size of architectural door, price and suppliers of valves in hydraulics and many others can automatically be grouped for any particular drawing. Such vital information can be extracted from the *graphic* block and automatically converted to *non-graphic* information such as spreadsheets to produce bills of quantities. Bills of quantities are vital when the design goes into production to assist the purchasing department to source the required materials. The proper management of component parts of any engineering drawing, in this case, blocks and attributes is another way in which productivity and throughput of complete engineering drawings with their associated information can be

DOI: 10.1201/9781003288626-7

accomplished, thus further justifying investment in CAD systems. The chapter also includes a practical application in the form of a complete case study presented in a tutorial format and worked-out examples to enhance users' skills in using AutoCAD.

7.2 CREATING AND SAVING BLOCKS IN CATALOGUES

Blocks or symbols are merely drawings that consist of various primitives such as lines, arcs, circles, etc. as described earlier. The procedure for creating them is therefore the same as any other drawing except that difference is in how they are saved and stored. Using the drawing principles and procedures described in Chapters 2 and 3 and before saving the display, there are two ways in which the block can be created and stored. The BLOCK command can be used to create temporary blocks that will only be available in the current drawing but cannot be used in other drawings. On the other hand, the WBLOCK command allows the user *Write* or save the block to a folder or chosen catalogue. The dialogue boxes for the two commands are shown in Figure 7.1. As can be seen from the figure, there are several parameters that need to be defined before a block can be created. The difference is illustrated by the need to specify a folder to save the block under the Write Block dialogue.

For each block that is created, several parameters need to be defined as shown in Figure 7.1 before a block can be defined/stored temporarily in the case of BLOCK command or saved onto an appropriate folder in the case of the WBLOCK command. The critical ones include the entities that make up the block, which can be selected using the **Select Objects** radio button on either dialogue boxes and the base point, also selected by clicking on the **Pick Point** radio button or defining and entering the absolute coordinates for that base point. It is crucial to select an appropriate base point as this will become the insertion point when inserting the blocks into other drawings. In this case, the object snap modes should be activated so that the base point can be the endpoint, centre, midpoint, etc. of one of the entities that make up the block. This will make it easier and more precise to insert the same block into a different drawing. Once all the parameters have been set, the block is created and stored in the memory of the current drawing in the case of the Block Definition dialogue box or saved into the user's preferred folder/directory or catalogue that can easily be remembered for retrieval and future use in other drawings. Typical examples of base points which will eventually become insertion points for examples of typical blocks used in different engineering disciplines are provided in Section 7.4.

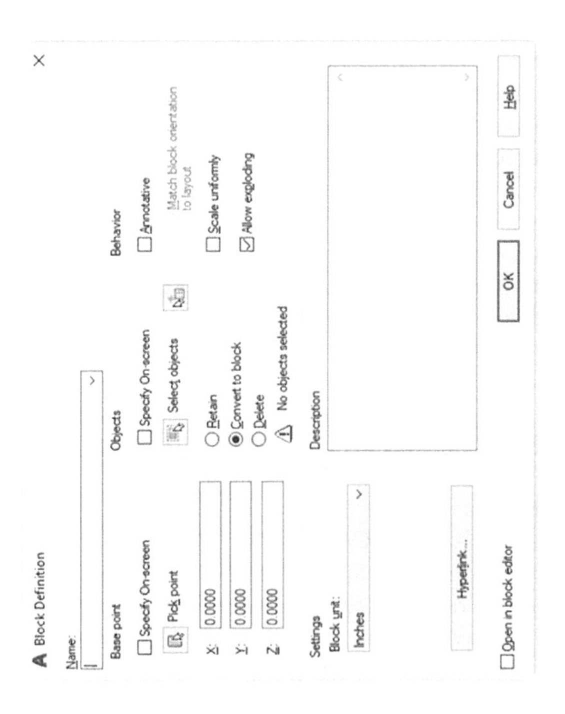

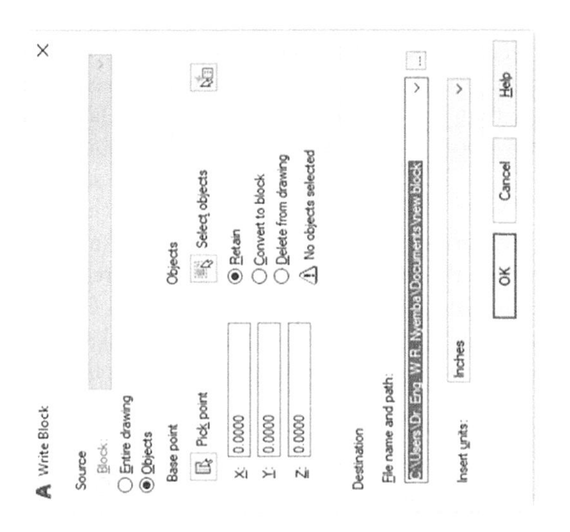

Figure 7.1 BLOCK and WBLOCK dialogue boxes.

7.3 INSERTING BLOCKS INTO DRAWINGS

Inserting blocks that would have been created either by the BLOCK or WBLOCK command is as straightforward as it was to create them. However, it should be noted that only blocks created using the latter command can be saved into an appropriate catalogue from where they can be inserted. Those created using the BLOCK command can only be stored and retrieved in the current drawing. There are also two ways in which created blocks can be inserted into current drawings. The INSERT command brings up a dialogue box in which the user can specify or retrieve recent blocks, blocks saved in the cloud or blocks from the current session. On the other hand, the DDINSERT command allows users to retrieve blocks from library catalogues that they would have created and save the blocks using the WBLOCK command.

What is critical to note are the parameters that need to be specified or defined before the block can be inserted into the drawing. The first is the insertion point, which should coincide with the base point defined when the block was created. The second is the scale factor, in which the stored block can be scaled up or down depending on what size will be required in the current drawing. Not all blocks saved will be inserted in the same orientation as they were created, hence the rotation angle in which the user can rotate the block to be positioned in the desired orientation. The two dialogue boxes for INSERT and DDINSERT are shown in Figure 7.2 displaying which and where the various parameters can be set.

However, it should be noted that when dealing with the creation and retrieval of blocks in AutoCAD, it might be best to utilize the mouse for selecting the base points or insertion points in conjunction with the object snap modes for precision, otherwise making use of absolute coordinates can be time-consuming and may lead to some errors. Blocks inserted into drawings are compound entities, comprising the entities used in creating them. In order to make any modifications to them, the user will need to split them into their individual elements using the **Explode** option or simply explode them after they are inserted.

7.4 TYPICAL BLOCKS AND THEIR ATTRIBUTES

Regardless of which field of engineering, all drawings consist of standard symbols that can be created in CAD and stored as blocks for use in future drawings thereby avoiding recreating them. Such symbols may be similar in configuration but different in size. For example, in the case of an M $d \times l$ bolt, there could be a wide range of bolts with different diameters (d) and lengths of shanks (l). In creating such blocks in AutoCAD, it is not necessary to create the whole range of bolts but only one, and when a different size is required in a particular drawing, it can be scaled up or down and rotated to position it accordingly in the drawing.

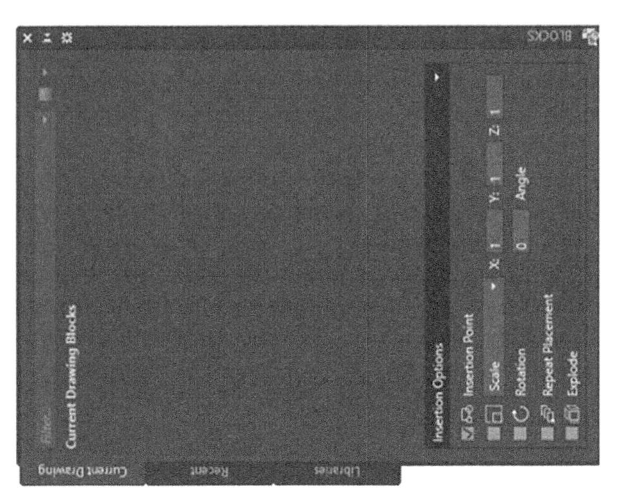

Figure 7.2 Retrieving and inserting blocks into drawings.

Table 7.1 Typical engineering blocks with base/insertion points

Symbol description	*Possible block name*	*Base/insertion point*	*Discipline/usage*
Single Door	SDOOR.DWG		Architecture
Variable Capacitor	VCAPACITOR.DWG		Electronics
Screw	SCREW.DWG	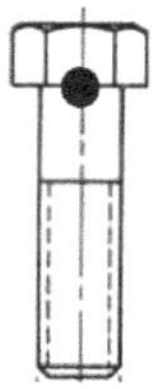	Mechanical Design
Valve	VALVE.DWG	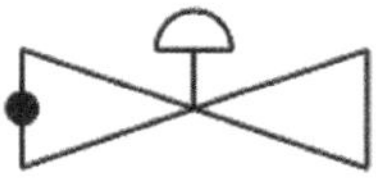	Hydraulics
Electrical Socket	ESOCKET		Electrical Design

Typical and standard engineering symbols that can be saved as blocks, together with suggested base/insertion points (black dot) are shown in Table 7.1. These are just examples and by no means exhaustive as there are several other symbols under any of the disciplines listed in Figure 7.1. The user can select an appropriate orientation and base point for storage and when inserting the block into a drawing, an insertion point is also selected following which the block can be rotated and scaled to a suitable final position. For example, a single door facing upwards can be rotated to face downwards or in any other direction.

Attributes of a block are its properties or labels that are often used to fully define what a block really is. These include such things as the name, label, price, special comment, supplier etc. The attributes are part of the drawing or block entity that is designed to hold text and link to graphic objects in CAD. Specific examples for typical blocks listed in Table 7.1 could be:

Single Door: fire rated, price, supplier, wooden or steel, etc.
Variable Capacitor: rating, price, supplier, purpose, etc.

Screw: diameter, head profile, length of shank, price, supplier, etc.
Valve: type, fluid, price, supplier, galvanized, plastic or brass, etc.
Electrical Socket: rating (13/15 Amp), price, supplier, 1 or 2 gang etc.

The desired name or label for the block can also be included. For ease of reference and retrieval, the same name chosen for the block can be used to avoid mix-ups.

7.5 APPLICATIONS AND TUTORIAL IN ELECTRONICS

To bring into perspective, the application and practical use of blocks, defining, attaching attributes, modifying attribute values, extracting attributes from a drawing and generating bills of materials will be better demonstrated through a worked-out example or tutorial. This example is derived from Electronics, where the purpose is to develop a general-purpose electronic circuit diagram for a printed circuit board (PCB) comprising the wiring diagram shown in Figure 7.3 and six electronic components that will be created and stored as blocks (Figure 7.4) before inserting them into the main wiring diagram, followed by attribute definition and extraction. The black dot on

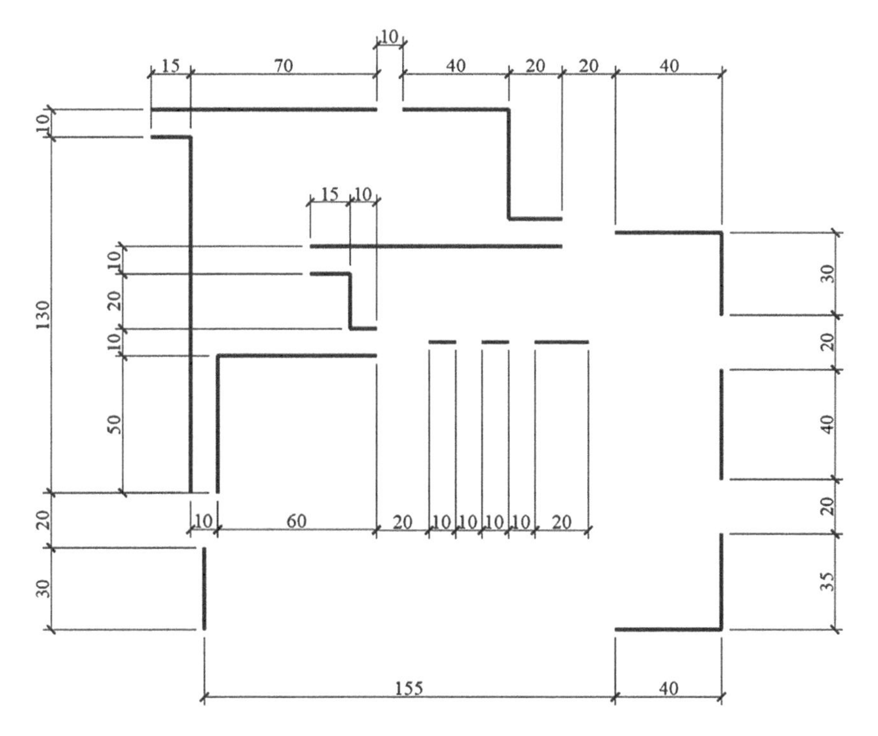

Figure 7.3 Main wiring diagram (shell) for the PCB.

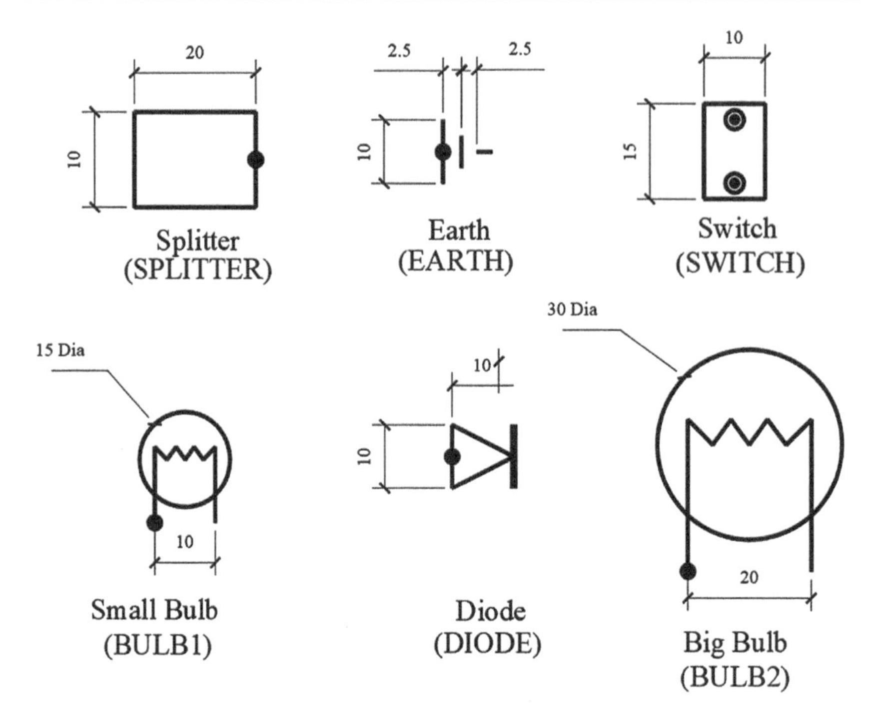

Figure 7.4 Electronic components (blocks) for the PCB.

each of the blocks is a guide on the best possible base/insertion point and the dimensions given on both figures are also a guide to enable users to go through this tutorial but are not necessarily typical measurements for PCBs. The attributes will be extracted from the drawing (graphic information) and used to automatically create a bill of quantities in the form of a spreadsheet, for use in ordering these components in a typical electronics company.

The first step would be to create the six electronic components as guided by the dimensions in Figure 7.4, then save them individually as blocks, using the capitalized names. The base point (black dot) in each of the blocks does not need to be drawn but is just an indication of where this could be best placed. However, users can choose any other base point on the blocks as long as it is appropriate to coincide with a definite insertion point on the main wiring diagram. The next step would be to create a layer, for example, wiring with an appropriate colour, where the main wiring diagram is provided as a shell as shown in Figure 7.3 and save this as a drawing (e.g., PCB.dwg). Having created the main wiring diagram and saved the six blocks, the next step would be to create another layer, with a different colour and then insert these blocks appropriately in the drawing as shown in Figure 7.5. It should be noted that some of these blocks are not necessarily oriented in the same manner as they were created but need to be rotated accordingly in order to be displayed as shown in Figure 7.5. For simplicity, the dimensions

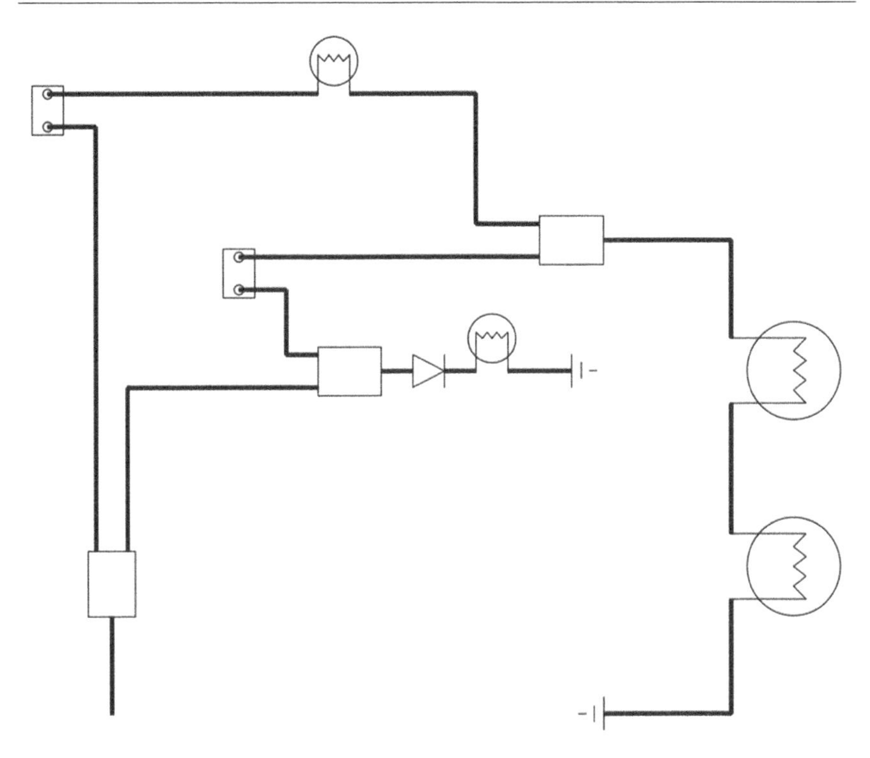

Figure 7.5 Complete wiring diagram for the PCB with blocks inserted.

and sizes used in creating the blocks should be maintained, hence no scaling, although, for the bulbs, scale could be used in which case only 1 bulb is necessary.

7.5.1 Block attribute definition, attaching and modification

Having completed the insertion of the six components as shown in Figure 7.5, the next step would be to define the attributes for each of the blocks and attach this to the blocks (non-graphic information attached to graphic information). To define attributes for each of the inserted blocks, the command DDATTDEF is used and this brings up the dialogue box as shown in Figure 7.6. Several parameters need to be defined on this dialogue box before the attributes can be attached. These are as follows:

Invisible: Turning on or off the attribute display on drawing
Constant: The attribute has a fixed value for all insertions of the block
Verify: Prompted to verify that the default attribute value is correct when inserting a block
Preset: Attribute set to a default value with no user entry
Tag: Heading or Field name of an Attribute

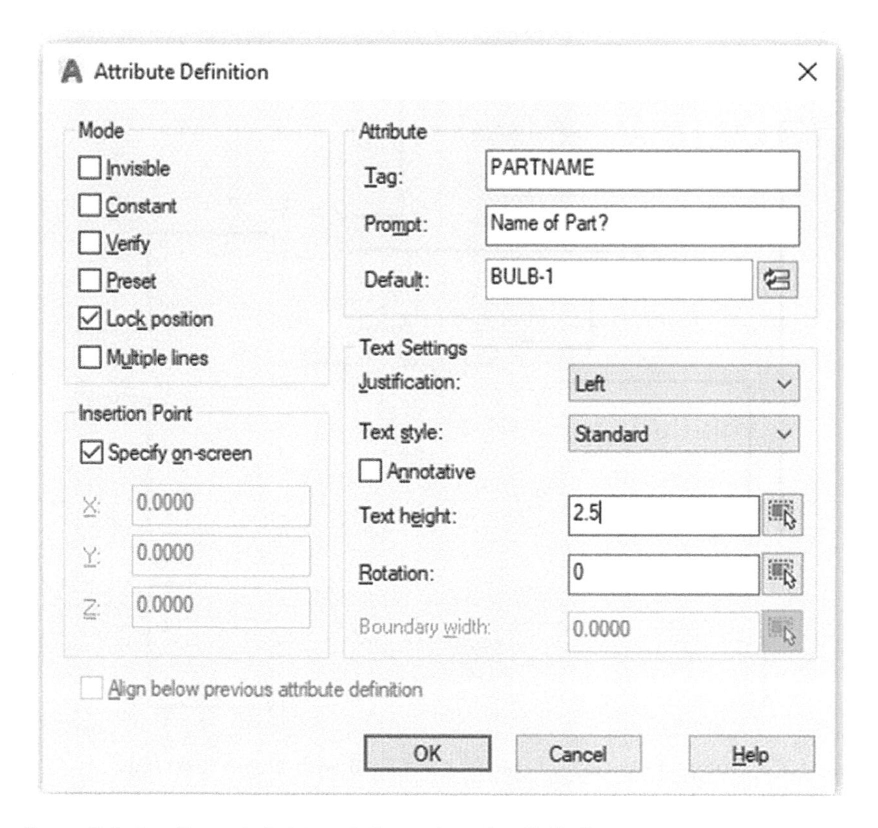

Figure 7.6 Attribute definition dialogue box for Bulb-1.

Prompt: Text that AutoCAD asks the user to enter a value of an attribute
Value: Default Value for an Attribute
Insertion Point: Corresponds to the base point where the attribute can be inserted and this can be selected using the mouse or by specifying absolute coordinates.

The attributes for the first bulb (Bulb-1) at the top can be defined as shown in Figure 7.6 by ticking **Verify** and the four attributes mentioned below; suitable height for the text in relation to the dimensions used (in this case, 2.5 as inserted in Figure 7.6), **Insertion Point** is best specified by the option 'Specify on Screen' in relation to the base point of the block. Accepting these parameters will prompt the user to select a suitable position close to the bulb to attach the attributes to the drawing. Figure 7.6 shows the attribute definition for PARTNAME. The same procedure can be repeated for the other three attributes of the bulb as shown in Table 7.2. For the last one, supplier, tick **Invisible** and **Constant** to automatically grey out **Prompt** and **Verify**. The four attributes should then appear near the bulb block as shown in Figure 7.7

Table 7.2 Attribute definition for the first bulb

Tag	*Prompt*	*Value*	*Insertion point*
PARTNAME	Name of Part?	Bulb-1	Specify on screen close to first bulb
RATING	Current Rating?	5 Amp	Align below previous attribute
COST	Unit Price?	$3.55	Align below previous attribute
SUPPLIER		Electrosales	Align below previous attribute

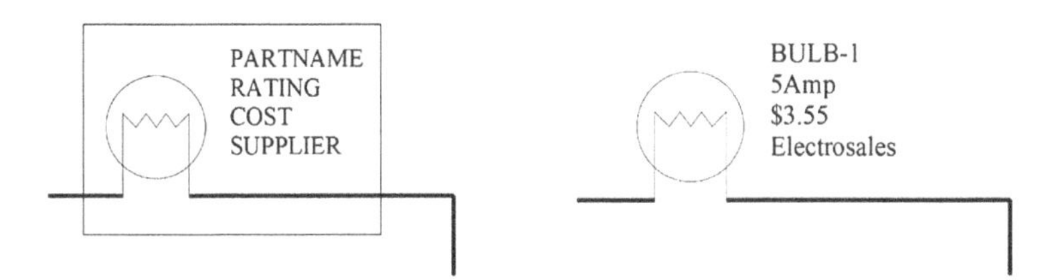

Figure 7.7 Appearance of bulb block attached with attributes.

In case of any errors, attributes of blocks can be modified at any time using the ATTEDIT command. To add the attributes to an existing block, the BLOCK command is used then followed by entering BULB-1 as the filename for the block. The bulb block is then selected together with the four attributes enclosed in a window as shown on the left of Figure 7.7, while maintaining the original insertion point. At the point of saving the block, it will prompt that the Block already exists, but this can be accepted to redefine the block and automatically, the attribute values are then displayed as shown in the second snapshot in Figure 7.7.

7.5.2 Managing block attributes

Various commands are associated with and can be used to manage blocks and their attributes when necessary. ATTDEF was used to define attributes for blocks, while ATTEDIT allows users to edit attributes independently of the blocks with which they are associated. DDATTE allows users to change the values of attributes on a block and ATTDISP is used to display (ON) or switch off (OFF) attributes in a drawing to avoid cluttering the drawing with too many details. On inserting blocks, users will be prompted to set any values if ATTREQ is set to 1(default) otherwise such prompts can be disabled by setting the ATTREQ to 0.

ATTDIA is also a system variable that can be used to control preset values coming out for verification on the command prompt, in which if it is set to 0, AutoCAD prompts for attribute values and with 1, no prompts are issued but a dialogue box is used instead. Multiple blocks in an array (rows and columns) can also be inserted using the MINSERT command. Blocks with different scale factors can also be inserted using the DDINSERT command, although it is generally discouraged and using the same scale in creating the block for inserting is preferred. If it is really necessary to scale a block such as when there are different sizes, for example, bolts and nuts, it is advisable to select **Uniform Scale** for all coordinates (x, y, z) to ensure that the block is not distorted in terms of its aspect ratio. The usual Copy (CP) command can also be used to insert duplicate blocks in other parts of the drawing to avoid going through the same process of inserting. In this case, the copied block will carry with it the attributes that may be attached to it.

7.5.3 Extracting attributes and generating bills of materials

The main purpose for adding attributes and their values to blocks is to be able to define components of a drawing adequately but more so, to be able to extract such non-graphic information and manage or organize it in such a way as may be necessary for a spreadsheet format for stocking, purchasing or selling. Such data are in the American Standard Code for Information Interchange (ASCII) format and can be extracted.

Non-graphic data (text) attached to blocks are in ASCII format and can be extracted and used in other platforms such as word processing, spreadsheets and other databases. Typical and practical use of such non-graphic information is in the generation of bills of materials for the costing of products in manufacture, quantity surveying in architecture and parts list from stores. The such extracted information will be more accurate and can be generated more efficiently than preparing them manually. Any changes to drawings will however require re-extraction.

Before extracting non-graphic data from a drawing there is a need to create a template file using such platforms as Notepad, to define how the extracted data should be organized. For the PCB tutorial, the following specifications can be typed into Notepad and saved in an appropriate folder, for example, PCBTEMPLATE.TXT. The attributes will be specified whether they are Characters (C) or Numeric (N) along with field length and number of decimal places.

PARTNAME C020000	Character with length of 20 and no decimal spaces
RATING C006000	Character with length of 6 and no decimal spaces
COST C006002	Character with length of 6 and 2 decimal spaces
SUPPLIER C020000	Character with length of 20 and no decimal spaces

After saving the template file in an appropriate folder, return to the AutoCAD drawing containing the blocks with attributes, then use the DDATTEXT command to extract the non-graphic data to a suitable text file. The Attribute Extraction Dialogue box is shown in Figure 7.8. Since tags and other fields in the PCBTEMPLATE.TXT file were separated by spaces, the Space Delimited File is ticked. The PCBTEMPLATE.TXT is selected after

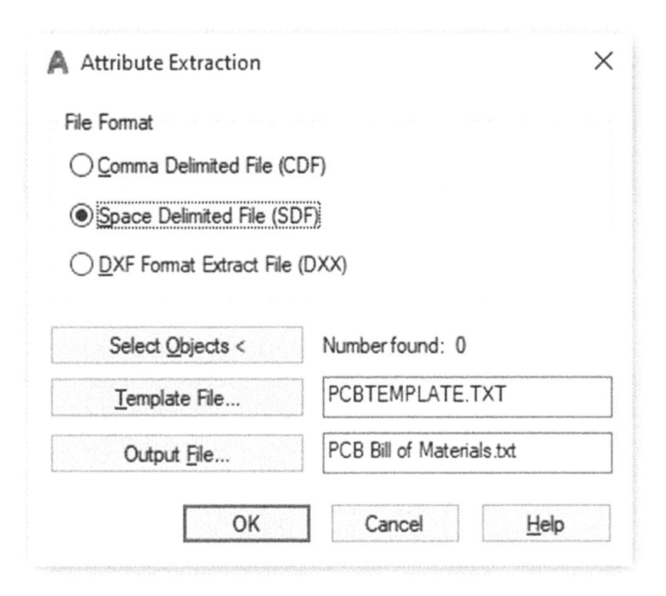

Figure 7.8 Attribute extraction dialogue box.

*PCB Bill of Materials.txt - Notepad

File Edit Format View Help

```
BULB-1          5Amp    $3.55   Electrosales
BULB-1          5Amp    $3.55   Electrosales
```

Figure 7.9 PCB bill of materials text file with bulb only.

clicking on the **Template File** on the dialogue box. A suitable name is then entered for the **Output File**, for example, PCB Bill of Materials.txt, followed by selecting ALL objects to automatically pick all attributes from the drawing. If the template file definitions do not match with the available attributes, an error will be prompted and the user can rectify this accordingly. The PCB Bill of Materials (BOM) can be opened from the folder where it was saved and should appear like that in Figure 7.9 with just one block (bulb) in it. The file created can be imported into Microsoft Excel or any other packages where it can be further refined to form a proper BOM or Parts List with appropriate titles, total costs etc.

The file can also be imported back into the drawing as a spreadsheet using the AutoLISP program, ASCTEXT.LSP, an aspect covered in the sequel and advanced book on CAD. The rest of the blocks on the PCB drawing can be created and attributes defined and extracted in a similar manner, following which the complete BOM can be generated, which is one of the exercises in Section 7.7.

7.6 CASE STUDY: CAD SUITE FOR THE AUTOMATIC GENERATION OF BILLS OF MATERIALS

The data, modelling and analysis in this section are a summarized version of a research/case study for the automatic generation of bills of materials within the same CAD suite environment, carried out at a motor vehicle manufacturing and assembly company (Nyemba & Lambu, 2006), which was subsequently published as a paper in the *Journal of Science, Engineering and Technology*, from where more details can be obtained.

A CAD facility in the form of an add-on menu for AutoCAD was developed for the automatic generation of bills of materials from CAD drawings under development and was implemented at a motor vehicle manufacturing company in Zimbabwe. Traditionally, most manufacturing companies' design and estimation functions are executed independently, invariably leading to errors or costly delays before completed drawings and estimations are passed on to production. This case study developed a module for automatically converting graphic information (drawings) to non-graphic information (spreadsheets).

Through the use of blocks and attributes as well as AutoCAD's high-level interface language, AutoLISP, ASCII programming and the Dialogue Control Language (DCL), an add-on facility was developed and customized (as explained in detail in Chapter 11) leading to the BOM module for the company. The module consisted of a user interface that incorporated a BOM pull-down menu, a dialogue-box-driven user input interface as well as online help. The implementation of the BOM module at the company resulted in the improvement and timeous creation of bills of materials by ensuring that all user input was carried out within the same CAD environment where the drawings were generated, to avoid any unnecessary duplication of tasks and thereby speeding up the delivery of such information to production. This also improved the efficiency of the company's drafting and estimation functions thereby minimizing errors that traditionally occurred when the two functions were carried out separately. Traditionally, companies with CAD facilities in industrializing countries produced detailed technical drawings in the design office and then passed these to estimators who prepare spreadsheets of bills of materials for costing before passing them to production. Such companies also traditionally tend to use non-graphic information in the form of spreadsheets in their factories as opposed to generated CAD drawings probably because the spreadsheets are much easier to interpret whereas detailed technical drawings can only be understood by qualified engineers and technicians.

Global competition and rapid changes in technology have forced engineers to adopt CAD and CAM systems for their everyday use, particularly in engineering and manufacturing. This has also been prompted by global reductions and shortages in 'hard' skills such as draftsmen, machinists, fitters, etc. as a result of reduced apprenticeship training and increased emphasis on quality of products and services through automation and integration of engineering design and manufacturing (Nyemba et al., 2021).

The use of CAD systems as the fundamental and basic method of producing detailed technical drawings has superseded manual methods of using the drawing board even in industrializing countries such as Zimbabwe. To cope with the demands of the Fourth Industrial Revolution and to prepare for the Fifth Industrial Revolution, more and more companies have invested in state-of-the-art computer systems including CAD and CAM for efficiency and in order to have full integration of the design and manufacturing functions. During the execution of this research, it was also established that most companies with CAD packages use them exclusively for the production of detailed technical drawings but little else with regards to modelling and analysis, hence the underutilization and thus not realizing maximum returns on their investments (Nyemba et al., 2021).

In addition, and although most CAD systems such as AutoCAD can be customized to provide avenues for integration, it requires a certain amount of programming skills and training. Most CAD systems in industrializing countries are operated by technicians such as draftsmen with some guidance from experienced engineers. However, such draftsmen are not skilled enough to utilize the vast potentials that CAD systems offer through programming with platforms such as AutoLISP, ASCII or DCL. Information gathered during the research also revealed that there was a need to develop user-friendly software modules to enable designers and draftsmen to automatically generate non-graphic information such as spreadsheets of bills of materials from graphic data such as drawings.

The module also needed to be compatible with the widely used CAD systems such as AutoCAD or for use by personnel with little or no programming experience. There was therefore a need for the integration of the inherent capabilities of AutoCAD with those of other widely used databases such as Microsoft Access and Excel, within and outside engineering design departments and other departments such as estimation and production, in such a way that information could be shared and exchanged more efficiently.

This was the basis on which this case study was carried out, demonstrated and tested on the automatic generation of bills of materials to enable engineering designers and technicians' access to online information on available stocks, suppliers, cost, manufacturer specifications or any other relevant details for achieving consistency of information within the company. Although the work was carried out as a case study at a motor vehicle manufacturing and assembly plant, and the add-on facility was tested in-house; it was done in a generic form to enable other design and manufacturing companies to adopt the same by simply adjusting a few elements of the programming code.

7.6.1 Database management in engineering design and manufacture

A database is normally referred to as a computerized record-keeping system that can store information and details about many different types of objects as well as how the information of objects is related to one another. Hence, database systems are comprised of data, hardware, software and users. The database management system (DBMS) is the computer software or interface that provides the link between the user and the actual database. Ideally, for efficient design and manufacture, there should be a seamless link and integration between the planning/design office with production as well as possible interconnections between the various functions under each of the units within the design or production. Such integration can be accomplished using external database commands such as those in AutoCAD. These include the AutoCAD Structured Query Language (SQL) Extension (ASE) where integration can also be achieved by developing appropriate software modules that interface with AutoCAD to enable users to link graphic and non-graphic information through AutoLISP or the AutoCAD Development System (ADS) in conjunction with the DCL. SQL was developed by an IBM research team in the mid-70s to facilitate data definition, manipulation and control using SQL statements (Alwan & Younis, 2017).

The AutoCAD SQL, just like many other database management software consists of an interactive and application programming interface. In the former, the user is provided with an immediate response, whereas with the application programming interface, the SQL statements are embedded in an application program and the user does not need prior knowledge of SQL to be able to query the database. However, in AutoCAD, the interactive interface is commonly used because the AutoCAD SQL does not support the embedded SQL (Lin & Hu, 2012). The AutoCAD SQL Interface (ASI) is a programming interface with SQL support, which enables users to access databases outside AutoCAD, whereas the AutoCAD SQL Link Interface (ASI Link) interfaces AutoCAD entities with external databases and the two form the ASE programming interface for developing AutoCAD and AutoLISP database applications (Lin & Hu, 2012).

AutoLISP is a high-level interface programming language for AutoCAD, which is a derivative of the list processing (LISP) programming language primarily used for Artificial Intelligence research. AutoLISP enables users to code macro-programs, routines and functions that are well suited for graphics applications, allowing them to access databases outside the AutoCAD environment (Nyemba, 1999), thus suitable for the development of the required module for linking AutoCAD with external platforms.

In comparison to other facilities available for customization in AutoCAD, as detailed in Chapter 11, AutoLISP is well suited for prototyping and for applications where controlling development and maintenance costs is more important than performance. The ADS is a programming environment developed using the C Language environment for developing AutoCAD applications. Although the C Language is the base on which ADS applications were developed, they are more or less similar in syntax and structure to functions written using AutoLISP, hence they are treated as external functions and are thus loaded using an AutoLISP interpreter.

Essentially, ADS and AutoLISP are similar but ADS applications are generally more efficient in terms of speed and memory usage due to the fact that ADS is compiled rather than interpreted or evaluated (Nyemba, 1999). On the other hand, ADS applications are capable of accessing certain facilities that AutoLISP cannot, such as the operating system and hardware, thus making them more suitable for applications that need considerable computation or interaction with the host environment. However, they can be time-consuming and expensive to develop (Lin & Hu, 2012).

The DCL is one of the most common and effective ways of getting user input through programmable dialogue boxes (PDB). Dialogue boxes, as shown in Chapter 11, are developed using ASCII files coded using DCL, which determines the arrangement of elements of the dialogue box (sometimes referred to as tiles, such as edit boxes, buttons etc.) as well as the size and functionality of each element. Whereas the parts of a dialogue box define how it is structured or used; the use and functionality of a dialogue box depend on the application that uses it. AutoLISP and ADS provide the functions for controlling dialogue boxes, although these are developed using DCL. In this case study, the normal blocks and attributes approach, as discussed earlier in this chapter, was adopted by developing interface programs in AutoLISP aided by DCL.

These enable users to pick from a customized menu, a facility that was employed to guide the users through the process of attribute definition and editing. In doing so, users do not need to leave the AutoCAD environment and so bills of materials can be generated or updated as the drawing is in progress, thus fewer or no omissions or errors. Apart from AutoLISP, ADS, ASCII and DCL, there were other considerations that were taken into account to develop a fully integrated system for AutoCAD to interface with external databases to avoid certain concerns and dangers that may arise. Such considerations included:

- Data security and control: Information must only be accessed by authorized personnel.
- Isolation levels: How do data in one transaction affect that in another transaction?
- Backup: What happens when the computer storage system fails, can data be retrieved?
- Keeping track of external databases in addition to drawings requires extra care.

7.6.2 Case study company details

The research was initiated by engineers from a medium- to large-scale motor vehicle manufacturing enterprise that employs in excess of 200 permanent employees and several contract workers at its operations in Zimbabwe. The company's products range from mini (12–30 passengers) and maxi (60–80 passengers) buses, truck and van bodies as well as other products such as storage tanks and trailers. The maxi bus bodies are built on three different makes of chassis, Scania, Mercedes Benz or Erf while the mini buses are built on the Mazda T3500 chassis. On the other hand, the truck bodies are built as drop-side or grain-side on Renault, Hino or Mazda T3500 chassis and the van bodies are supplied as refrigerated, insulated or standard aluminium built on Hino, Mazda or Nissan chassis. The company's departments within the production plant include – machine shop, assembly lines, paint shop, polishing section as well as research and development.

In addition, there are also ancillary sections to augment these main sections and these include support services such as purchasing, drawing office, quality assurance, costing and estimation as well as marketing and sales. The research was based in the technical drawing office, which included the office of the estimator. Within this office, all designs of products are carried out, including the generation of bills of materials for the purposes of costing. The company generally produces as per orders from customers and in a few isolated cases, new products for general sales. Upon receipt of an order, the sales department raises a job request consisting of the customer's specifications and other relevant information for the manufacture of the product.

Upon receipt of an order and depending on whether the requested product is entirely new or not, the technical office would either generate a new drawing or modify an existing one to suit the customer's specification and order. Before 1997, the company produced their detailed technical drawings manually, using drawing boards, large-size papers and numerous filing cabinets. Since the turn of the new millennium, all their systems have been gradually computerized in virtually all departments including the technical office where they adopted AutoCAD as their CAD system of choice. At the time of carrying out the case study, the generation of bills of materials and estimation were still largely manual, but derived from the CAD-generated drawings. The bills of materials were basically in three forms, that is, cutting, bulk or bought-out lists. However, the three forms were complementary and formed the complete materials requirement specification for each product. Of prime concern to the engineers and management was the need for an efficient system for generating bills of materials from the technical office. This was one of their bottlenecks and thus they needed a system that would enable them to timeously and easily create bills of materials while avoiding the repetition of tasks but maintaining some integrity and accuracy in the spreadsheets.

Somehow, at that time, they felt, rightly so, that with the advent of computers and rapid changes in technology, there must be a way of avoiding duplication of tasks, where estimators would need time to study and understand detailed technical drawings before they can manually generate an accurate spreadsheet of the materials required and their costs. This was prone to errors and invariably resulted in a back-and-forth process to avoid over or under-quoting a customer. Typical errors also included omissions which resulted in considerable delays in the delivery of ordered products. In today's competitive world, especially with new and rapid changes in technology, such a company could end up losing a contract, especially if they present revised quotes to customers. This may cost them their reputation and future contracts, thus the need for accuracy and short lead times.

The major requirements for the utility were therefore outlined as follows:

- The system must allow users to associate graphic objects (drawings) with non-graphic information (spreadsheets and databases).
- The non-graphic information associated with the drawing objects must be entered by the user within the AutoCAD environment to avoid duplication of tasks.
- The system must, as far as possible, guide the users through the process of linking drawing and database information through prompts and automated tasks.
- There must be a facility to extract all non-graphic information from a drawing, and export the information to a separate file where it can later be retrieved and stored in an external database or used for the purposes of costing.
- As much as possible, the utility must allow all operations to be carried out within the AutoCAD environment to reduce the time required to produce a drawing as well as generate the associated bills of materials for the same.

7.6.3 Development of the BOM add-on utility

Two options were explored for linking drawings in AutoCAD with external databases. The ASE and the ADS required a considerable amount of programming experience on the part of the developers as well as the users. In addition, the use and availability of database software such as dBase IV or Access and the ability to operate these were basic minimum requirements to successfully use ASE and ADS. On the other hand, the use of blocks and attributes aided by AutoLISP and ASCII programming required little or no programming experience from the users, hence this was chosen as the most suitable option. Requests for modifications and upgrades of the software can also be requested from the developers, in the same manner that new versions of the software are developed.

As outlined earlier in this chapter, the use of blocks and attributes in AutoCAD requires that the user defines all these after a drawing is completed. Somehow, this defeats the whole purpose of the research in that some omissions can still be encountered. The focus was therefore to ensure that after creating a part of a product, the blocks and attributes are then immediately defined before moving to the next part. That way, a bill of materials will automatically be generated as the drawing progresses, thus removing possibilities of errors of omission.

The definitions of the commands that performed all the functions in the BOM Utility that was developed were programmed using AutoLISP. Initially, the function definitions for each of the different tasks, the dialogue boxes and their associated functions were coded and tested. The online help facility was also included to make the whole system user-friendly. Finally, the menu files for the BOM utility were also coded and linked with the newly defined AutoCAD commands and added to the AutoCAD menu as shown in Figure 7.10.

Four AutoLISP programs were coded to facilitate the entry of non-graphic information within the drawing database. Selected programs are displayed in Appendix A1. The menu file, **bom_menu.mnu** (Appendix A1.1), created using ASCII programming in a standard text editor was loaded automatically when AutoCAD was started, using the AutoLISP file, **bom.lsp** (Appendix A1.2). This availed the BOM pull-down menu as shown in Figure 7.10. The file, **bom_entry.lsp** facilitated the entering of all attribute information into the drawing database through the Single Entry option of the BOM pull-down menu. After completing a drawing, the file, **compile_bom.lsp** (Appendix A1.3) compiled all attributes and their information in the drawing database and produced a table of the attributes for a particular product.

To facilitate the availability of the three AutoLISP programs in AutoCAD, the fourth file, **acad.lsp** (Appendix A1.4), was programmed in such a way that all three programs were automatically loaded each time AutoCAD was started. In the later or newer versions of AutoCAD, it is necessary to simply append the few lines in this program to an existing **acad.lsp**. Although the

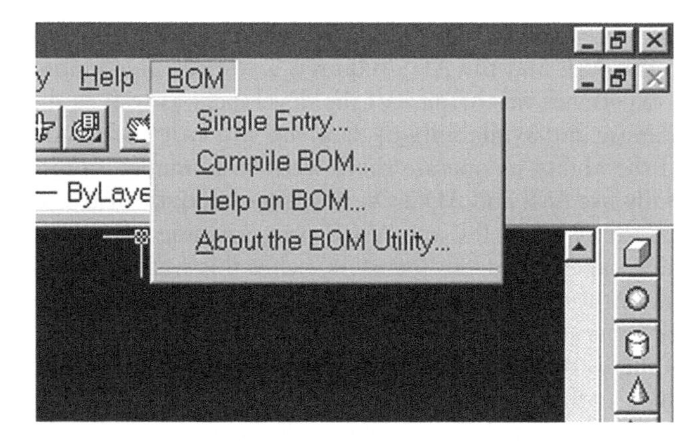

Figure 7.10 BOM utility pull-down menu created in AutoCAD.

initial idea of entering attributes on the command line reduced the chances of errors and omissions, a lot of input from the user was required especially in verifying attributes entered, hence the use of dialogue boxes as an indispensable tool for facilitating visualization, user interaction and data entry.

The use of dialogue boxes would eliminate the need to verify attributes entered, as is the norm but this can all be accomplished by clicking 'OK' to continue. Several dialogue boxes were developed through the coding of DCL programs as listed and coded in Appendices A1.5 and A1.6. These include dialogue boxes for entering attribute text, entering block names as well as attribute information as shown by the sample dialogue boxes in Figure 7.11. Attributes of components were entered using the dialogue boxes shown in this figure, through the two DCL programs – **attr_info.dcl** (Appendix A1.7) and **attr_info2.dcl**.

The attribute text properties dialogue box as shown in Figure 7.12 (left) was defined and derived from the DCL program, **att_txt_prop.dcl** (Appendix A1.5), while each component was defined and identified as a block and the block name entered using the dialogue box created by the DCL program, **blk_name.dcl** (Appendix A1.6). Through the AutoCAD customization facility as detailed in Chapter 11, an online help facility was also developed as it was evident that users preferred such a facility as opposed to the use of manuals and guidelines. Therefore, to ensure the add-on utility was user-friendly, the online help file, **bom_help.ahp** was coded using ASCII and

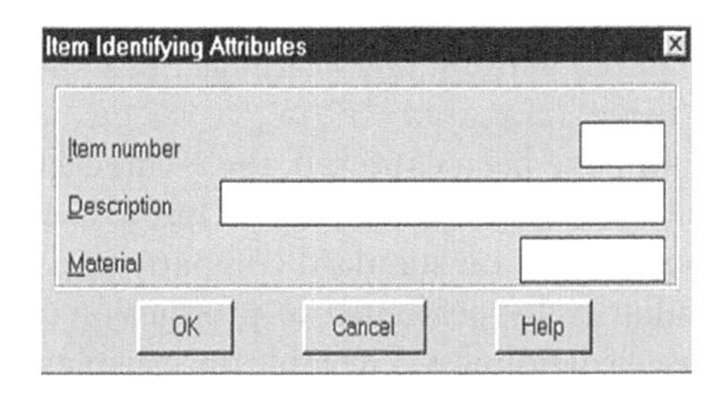

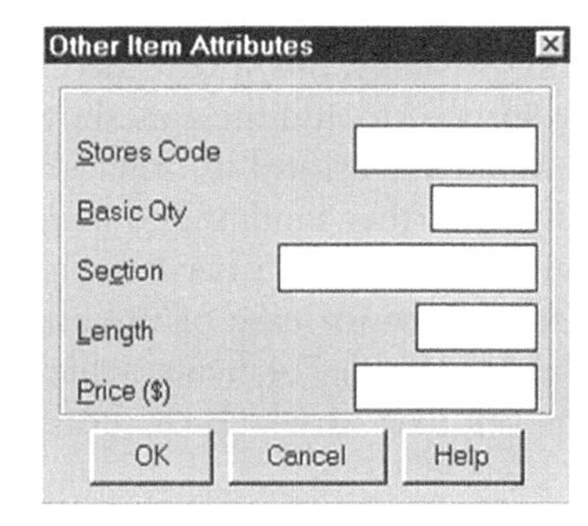

Figure 7.11 Dialogue boxes for entering attribute information.

Figure 7.12 Dialogue box for attribute properties and sample BOM help.

availed under BOM Utility pull-down menu. Figure 7.12 (right) shows a snapshot of the general help under the BOM utility while the 'Help' button on the individual dialogue boxes provides specific help on how to use each dialogue box.

7.6.4 Implementation of the BOM utility

As the drawing of a component or assembly is in progress, the attributes and properties are also continually being entered, following which a bill of materials spreadsheet can be extracted. After the extraction of attributes to a specified text file, the information can then be imported into any of the commonly used spreadsheet packages such as Microsoft Excel using several different options such as Comma Delimited File, Space Delimited File or the Data Exchange Format. However, the Comma Delimited File option gives the user more control over what is produced by the extraction.

The resulting text file consists of a number of specified rows and columns depending on how the template output file is defined, a typical one of which is shown in Appendix A1.8. The template output file also determines field lengths permitted for each column or whether or not the value expected from the user should be character or numeric, as explained earlier in Section 7.5.3. Table 7.3 shows a typical text file that was derived after extraction and generated using the BOM utility for the Aluminium Bonded Standard Van Body, containing information such as quantity, description of component, material or form for the component, cut size and cost in the local currency. Usually, this spreadsheet is generated without any headings but it can be spruced to add these as shown in Table 7.3.

While the critical information would have been captured, the created file can be further modified to include headings as described above but also to suite the company's needs such as displaying it on standard company paper with a logo for ease of use and familiarity by personnel in production as shown in Table 7.4, hence achieving user-friendliness. The table only displays the top part of which the bottom part will comprise what is in Table 7.3.

Table 7.3 Typical output text file after importing from AutoCAD

1	4	Floor plate	MSP	2.5 × 1200 × 2360	23,600
2	4	Floor Plate	MSP	2.5 × 1200 × 1975	19,750
3	14	Cross Member	RSC	76 × 38 × 2163	27,040
4	4	Chassis Runner	MSP	4.5 × 232 × 2400	24,000
4a	4	Chassis Runner	MSP	4.5 × 232 × 1970	19,700
5	16	Side Pillar	TUBE	30 × 30 × 1.6 × 1855	9,275
6	4	Fr. Corner Post	MSP	2.5 × 311 × 1855	18,550
7	4	Rr. Corner Post	MSP	2.5 × 311 × 1855	18,550
8	2	Floor Plate	MSP	2.5 × 120 × 2360	23,600

Table 7.4 Customized bill of materials for the company

AUTOBODY MANUFACTURERS AND ASSEMBLY (PVT) LTD.					
BILL OF MATERIALS			**JOB No.**		2088-9
PRODUCT:					
Aluminium Bonded Standard Van Body on Mazda T3500			**DATE**: 3/11/99		**QTY**: 2
CUSTOMER:					
ABC Transporters (Pvt.) Ltd.			**DRG. No.**		E107-7
COST ACCOUNTING					
Item No.	**Quantity**	**Description**	**Material/ form**	**Cut size/section**	**COST/ ZWL$**
1	4	Floor Plate	MSP	2.5 × 1200 × 2360	23,600
2	4	Floor Plate	MSP	2.5 × 1200 × 1975	19,750
3	14	Cross Member	RSC	76 × 38 × 2163	27,040
4	4	Chassis Runner	MSP	4.5 × 232 × 2400	24,000
4a	4	Chassis Runner	MSP	4.5 × 232 × 1970	19,700
5	16	Side Pillar	TUBE	30 × 30 × 1.6 × 1855	9,275
6	4	Fr. Corner Post	MSP	2.5 × 311 × 1855	18,550
7	4	Rr. Corner Post	MSP	2.5 × 311 × 1855	18,550
8	2	Floor Plate	MSP	2.5 × 120 × 2360	23,600
TOTAL COST					**184,065**

From the generated output text file containing the list and specifications of materials used in a typical Mazda T3500 van body, the information can then be superimposed and inserted on the company's usual documents such as Bills of Materials Cost Accounting form for preparing quotations by the cost accountant or estimator. The ordinary text file is accessible or editable in Microsoft Excel or any other platform that handles forms. In addition, this can also be availed throughout the company through LAN. Table 7.4 shows a customized and typical form where the information from Table 7.3 was imported and superimposed on the company form. However, pseudonyms were used for confidentiality.

7.6.5 Implications and achievements of the BOM utility

Most engineering and manufacturing companies are involved in the production of detailed technical drawings from the design or technical offices to their production departments. In the process of creating such drawings, non-graphic information such as bills of materials are required and they can benefit from using the BOM utility that was developed at the motor vehicle manufacturing and assembling plant. The open architecture and customization capabilities of AutoCAD allow users to program and tailor-make the package to their specific needs. The **BOM Utility** was the collective name for the add-on facility developed and customized in AutoCAD for automating the generation of bills of materials while the drawing is in progress. The utility is accessible in all the Windows versions of AutoCAD. This additional menu enables users to navigate through several dialogue boxes while entering the necessary information in the definition of blocks or component parts of the assembly.

Once the drawing is completed the information entered is compiled and exported to a text file, which can then be imported into any spreadsheet package such as Microsoft Excel. The AutoCAD system file, **acad.lsp** was modified to automatically load all the routines that are needed to run the menu options at the start of an AutoCAD session. On the other hand, the **Help on BOM** provides the user with a help facility specifically developed for the BOM Utility. This provides the user with information and instructions on how to use the utility. A hardcopy user's manual on the BOM utility was also provided.

The integration of CAD with external databases enhances the consistency and flow of information within the company by exploring some of the customizable files provided with AutoCAD through the ASE, ADS, AutoLISP and ASCII Programming as well as the use of blocks and attributes. For this case study, blocks and attributes were chosen as it was easier to understand in combining drafting and estimation functions.

The use of ASE and ADS required good database software as well as properly qualified personnel, capable of programming in various languages. While it is ideal to have such a system to help companies minimize the number of errors and possible omissions when the bills of materials are generated, a number of issues are needed to be addressed for maximum and beneficial use of the BOM facility. These included the availability of a Local Area Network (LAN) within the company to facilitate the smooth flow and consistency of information within the company. The linking and connection of all departments ensured that information could be accessed from a central database. Staff within all departments also needed to be trained on the use of the BOM Utility system to avoid hiring external consultants. In addition, available software needed to be completely up to date and consistent throughout the company. The newer versions of AutoCAD as well as requisite database software were recommended.

Several advantages and benefits can be derived from the use of the BOM Utility such as automating the generation of bills of materials through linking AutoCAD with external databases, thus saving a lot in terms of disk space as well as duplicate files. The availability of information online also leads to fewer inconsistencies in data and hence more accurate records and improved efficiency in design. Documentation of design and production planning are easier to manage because information can be updated with ease thereby readily producing customer quotations timeously with reduced lead times.

There are also numerous possibilities for expanding the system to link with external and associated companies such as suppliers and customers. This could further reduce the lead time as customers could approve drawings and designs as well as acceptance of the quotations online as a stage-by-stage process. Similarly, suppliers can also timeously provide information and specifications for materials available in their stocks, including costs, in order to also timeously generate quotations required by customers.

Customization of AutoCAD was achieved through programming in AutoLISP to produce a BOM utility and ASCII for the production of the additional menu as an add-on facility to AutoCAD. DCL was used to refine the module by providing dialogue boxes that were user-interactive. A well-established LAN enhances the use of such a system as it improves the lead time for generating accurate quotations. Although the research was carried out at a motor vehicle manufacturing plant, the BOM Utility was developed in a generic form such that other engineering and manufacturing companies could also make use of the facility and benefit from the automatic generation of bills of materials. This was a typical demonstration of the capabilities of CAD software, customized and linked to external databases to enable the integration and extraction of critical non-graphic information such as specifications, material form, cut sizes and costs from graphic information such as a drawing or model in progress for the purposes of planning and preparing quotations for customers.

7.7 SUMMARY

Virtually, all engineering and technical drawings comprise a multiplicity of standard symbols that are used to represent components such as doors and windows in architecture, fasteners in mechanical design and diodes and capacitors in electronics. To avoid the repetitive nature of redrawing such common symbols in engineering drawings, blocks can be used to create and save these symbols for use in other drawings. In this chapter, various AutoCAD tools necessary to create, save and retrieve blocks from catalogues were outlined. In addition, attributes for blocks were defined, attached and displayed on the drawings, from where non-graphic information was extracted to automatically generate spreadsheets that are useful for costing products and preparing quotations in manufacture, quantity surveying in architecture, parts list from stores in a production factory, etc. Such automatically generated and extracted non-graphic information will naturally be more accurate and can be generated faster than preparing them manually. This was demonstrated in a case study for the automatic generation of bills of materials for a motor vehicle manufacturing company in

Zimbabwe where both graphic and non-graphic information was handled within the same CAD environment suite.

7.8 REVIEW EXERCISES

1. For five disciplines in Engineering and the Built Environment, list at least five typical symbols that are used in each of these fields that can benefit from their storage as blocks in CAD for later retrieval and usage in other drawings.
2. What are the most critical parameters that must be defined to enable one to create and save a block as well as to insert a block into an existing drawing?
3. There are several commands at the disposal of the AutoCAD user to adequately manage and organize blocks and their attributes. List at least five of these and for each and clearly explain what it is meant to achieve.
4. ATTREQ and ATTDIA are block commands that are used to set variable parameters in binary fashion (0 or 1). Explain what each of these system variables stands for and what it means if each is set to 0 or 1.
5. With reference to the PCB wiring diagram as shown in Figure 7.5:
   i. Use Table 7.5 below to attach all the attributes and values to all the blocks in this wiring diagram, then extract these to a temporary text file using Notepad.
   ii. Export the resultant temporary text file to Microsoft Excel to enable you to prepare and organize a proper PCB BOM, with appropriate column titles, total cost, etc. as shown in Table 7.5.

Table 7.5 Attributes and values for components in the PCB diagram

Description	*BLOCK NAME*	*RATING*	*SUPPLIER*	*COST*
Bulb	BULB-1	5 Amp	Electrosales	$3.55
Bulb	BULB-2	6 Amp	Electrosales	$6.90
Diode	DIODE33FS	15 Amp	Keltronics	$1.05
Earth	EARTH-1	0	Lucas Electric	$1.30
Earth	EARTH-2	0	Lucas Electric	$1.50
Switch	SWIT425	13 Amp	Keltronics	$2.30
Switch	SWIT450	15 Amp	Keltronics	$2,65
Splitter	SPLITTER	15 Amp	Lucas Electric	$0.75

Chapter 8

Three-dimensional wireframe and surface modelling

8.1 INTRODUCTION

The multitude of engineering designs and detailed technical drawings in common use around the world is two-dimensional, defined generally by the (x, y) coordinates only. These are either presented in orthographic or isometric (pictorial) or axonometric projections. These types of projections, particularly the orthographic ones contain details such as specifications and dimensions. Even though they contain so much detail, their use is also because of their simplicity in production compared to the stages required to produce 3D models. Nevertheless, while the 2D drawings are sufficient to provide information on site and in production plants, they are not very user-friendly for laymen in engineering or technical appreciation of projections and standards in specifications, more so for management who make decisions but by and large are non-technical personnel. By nature, and convention, 2D drawings do not contain information such as materials that make up a product, mass moments of inertia, surface details etc. unless these are provided as specifications on the 2D drawings. As such, analysis for strength and prediction of performance in use would be difficult to carry out on 2D orthographic or isometric projections.

Although 3D models are more involved in generating them in that, the details pertaining to the ultimate product are built and specified during development, this chapter provides guidelines on how to efficiently and quickly generate 3D models, and analyse or automatically convert them to 2D drawings. Apart from being pictorial and useful to non-technical personnel, 3D models are useful for analysis and determining aspects such as stresses and strains under loading and functioning as well as mass moments of inertia, weight and surfaces bounding them.

The visualization of objects in three dimensions is important even for non-technical personnel as it helps in viewing the object from different angles, thus fully understanding what a product should look like in real life, apart from being able to simulate in order to predict its performance. In addition to the (x, y) coordinates, to fully define or represent objects, a third dimension (z) is required, apart from then specifying materials and other

DOI: 10.1201/9781003288626-8

properties to fully define objects. Generally, objects in 3D modelling can be represented in three different ways: wireframe, surface and solid models and within these main categories are other subcategories, the content of which will be covered in this chapter, for wireframe and surface models while the next chapter will deal with solid modelling.

8.2 3D COORDINATE SYSTEM

The default display on AutoCAD screens is the normal Cartesian coordinate system with y positive upwards and x positive to the right, and the user coordinate system (UCS) icon is always located at the bottom left corner of the screen. However, this can be shifted as desired as will be seen later in this chapter. For the 3D coordinate system, the default (x, y) orientation is the same for 2D but the additional (z) coordinate is positive pointing towards the viewer and out of the screen. Similar to the 2D representations, the 3D UCS can also be changed and oriented in a desirable manner in order to capture the maximum display.

8.3 VIEWING POINTS

For maximum visualization, objects that are modelled in 3D need to be oriented appropriately for maximum view. The position of the observer's eye is critical to viewing an object and this is dictated by the viewing point, or more commonly the position of the eye in relation to the 3D UCS. The general rule is that when the observer's eye is placed directly behind a particular axis, then the coordinate for that axis will be negative, when the eye is positioned directly against an axis, the coordinate for that axis is positive and when the eye is positioned at right angles of any of the axis, then the coordinate for that axis will be zero. The eye position in relation to an axis and the resulting polarity is shown in Table 8.1.

Table 8.1 Polarity of viewing point in relation to eye position

Eye direction	*Axis*	*Polarity*
(Against an Axis) ◁	←	Negative
(Behind and Axis) ◁	→	Positive
(Perpendicular to an Axis) ◁	↑	Zero

For 2D orthographic projections, therefore when one starts AutoCAD, the default UCS icon has x positive going to the right, y negative going upwards and z positive towards the observer. As such, using the general rules in Table 8.1, the first view that the viewer sees on starting AutoCAD is the Plan, in which both the x and y coordinates are zeros because the observer looks at the screen, perpendicular to both the (x, y) axes. The z coordinate is positive and will have a value or magnitude, indicative of a scale away from the object. Usually, it is easier and convenient to work with binary digits (0 and 1) but any other numbers can still possibly be used. For the default start-up or Plan view, therefore, the viewing point is (0, 0, 1). The other 2D orthographic viewing points for the elevations are, Front Elevations (0, –1, 0), Right Side Elevation (1, 0, 0), Left Side Elevation (–1, 0, 0) etc.

To be able to visualize any object in three dimensions, that means each of the three coordinates must have a magnitude, positive or negative but must not be zero. Usually, the best orientation, depending on the object being modelled, is (1, –1, 1) to tilt the UCS icon as shown in Figure 8.1. The foregoing was more applicable to the earlier versions of AutoCAD but the more recent ones either make use of the World Coordinate System (WCS) at the top right of the graphics window by clicking and holding on to the LMB to spin it to the desired orientation or simply making use of the four preset 3D views, SW, SE, NE and NW isometric orientations accessible under the View pull-down menu. These four are usually sufficient to adequately display the 3D images in almost all possible angles but the user can set their own using WCS.

The other commonly employed viewing point for three-dimensional orientations are shown in Figure 8.2 but any other orientations can be used depending on the object being modelled.

Regardless of which viewing point is chosen, the appropriate UCS plane must be used to model such items as circles and ellipses. The various UCS

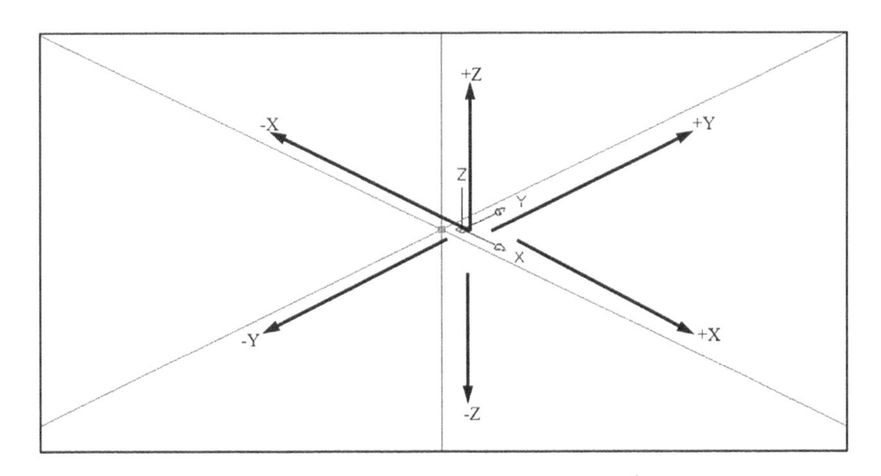

Figure 8.1 3D orientation of the UCS at viewing point (1, –1, 1).

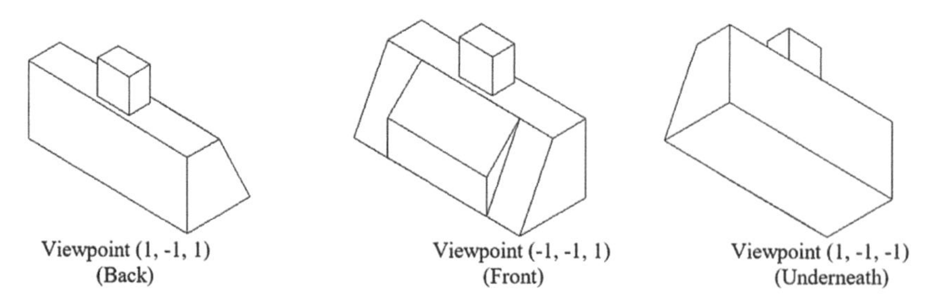

Figure 8.2 Commonly used 3D viewing points.

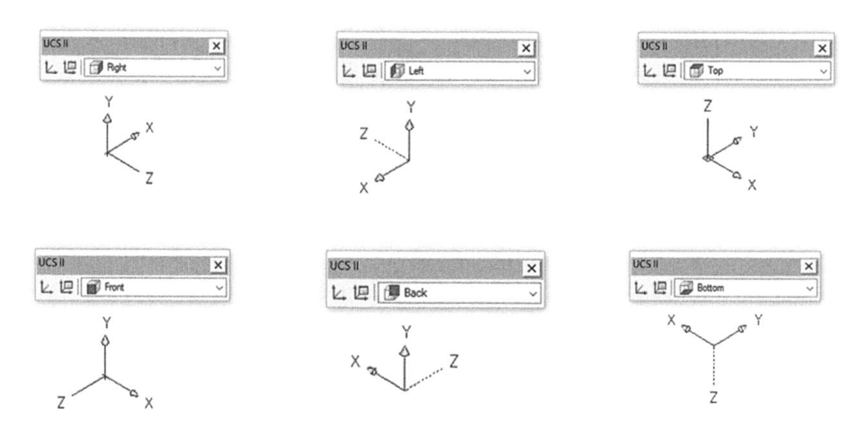

Figure 8.3 Orientations of the UCS icon.

planes are readily available by turning on the UCS II icon bar and using that to choose the appropriate plane to insert such objects. Usually, by default, the UCS II icon bar is turned off and this can be turned on by right-clicking on the empty space to the right and adjacent to the menu slide bar, then selecting AutoCAD and lastly UCS II to turn it on. The various orientations of the UCS icon are shown in Figure 8.3. The default plane for drawing any objects is the XY plane and users must therefore be careful to choose the right plane to insert entities such as circles and ellipses otherwise they may be distorted if the wrong plane is used.

8.4 WIREFRAME MODELLING

The basic and simplest representation of objects in three dimensions is the wireframe model, which is almost similar to the specifications and construction of 2D drawings in terms of absolute coordinates and lines except that in three dimensions, the specification of points contains a third (z) dimension. As such, the wireframe model depicts only the shape or 'skeleton' of the object but no other details. Although 3D wireframe modelling is fairly

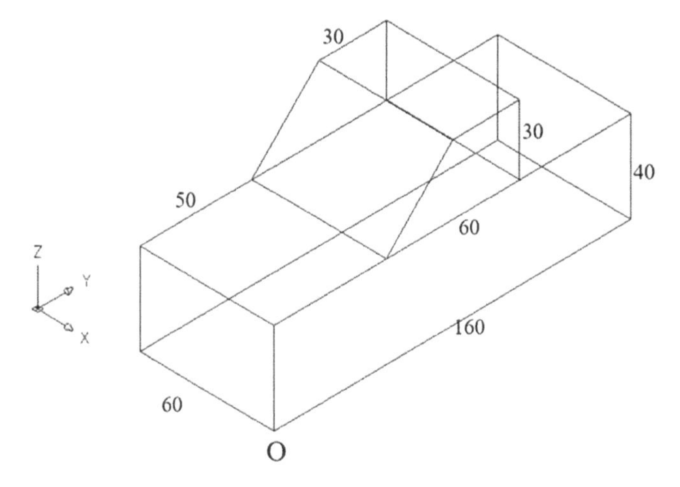

Figure 8.4 Illustration of a simple wireframe model.

straightforward to construct, it is not often used because of the numerous lines that cannot be hidden, thus sometimes creating confusion.

In this section, a simple wireframe model will be demonstrated to show how the line command is employed to generate wireframe models. It must also be noted that, as much as there are other ways of drawing lines from one point to another, these can be specified in the same manner as the construction of 2D drawings, that is, using *Absolute*, *Relative* or *Polar* coordinates or a mixture of any of them, except that the specifications should contain the third (z) dimension. Figure 8.4 shows an illustration of a simple wireframe model with the minimum number of dimensions required to construct it, placed along the measured edges.

In order to orient the model as development progresses, the viewing point can be set to a suitable 3D view such as (1, –1, 1) or spinning the WCS to a desirable angle. The 3D wireframe model can then be constructed using the line command, starting at any point such as the bottom point (O) nearest to the observer and moving in any direction as long as the correct specifications are made, *Absolute*, *Relative* or *Polar*, ensuring that the third (z) coordinate is included. For this type of construction, the relative coordinates option will be used as it is easier to manipulate the movements by specifying changes in any of the three axes. This would be the same as in 2D drafting where the changes (Δx, Δy) were specified except in this case (Δz) is also included for 3D.

Dimensions which are not provided in Figure 8.4 are not necessary as they can be worked out from those that are given. The line (L) command is initiated, starting at point O using relative coordinates (Δx, Δy, Δz) going towards the right at the bottom, the next point will be **@0, 160, 0** followed by the second 40 units' line going upwards **@0, 0, 40** back to the point above the start point, **@0, –160, 0** etc. all the way around the model until it is complete. Object snap modes such as endpoint, midpoint, intersection, centre etc. can also be employed in the construction where necessary to be precise and speed up the modelling.

Having completed the 3D wireframe model, users can spin it around to see views from different angles. This can be accomplished by using the WCS tool at the top right of the graphics window by clicking on it and holding the LMB as the user spins it to any desired orientation or makes use of the preset 3D views, SE, SW, NW and NE. However, unfortunately, that is about all that can be done with 3D wireframe models as they only depict the edges and 'skeleton' of an object. Even to try and experiment by hiding edges that are supposedly hidden, using the HIDE command, the same model will reappear. As such, 3D wireframe models are only useful for quickly constructing the 'skeleton' of an object but the more edges are available, the more confusing the wireframe model becomes as edges behind others cannot be hidden, hence the need for surface or solid modelling.

8.5 SURFACE MODELLING

Surface modelling is a mathematical technique used in computer-aided design applications for displaying solid-appearing objects (Maqsood et al., 2020). Surface modelling enables users to visualize objects at specific angles with solid surfaces. Surface modelling is a commonly used technique for architectural drafting and designs as well as rendering. This technique has a wide variety of applications such as in marine vehicles, body panels of automobiles and aircraft structures, as well as virtually all objects with surfaces (Maqsood et al., 2020).

In addition to holding information pertaining to the shape and edges of an object, surface models are also used to define the surfaces bounding the edges of objects. There are several ways in which surfaces of objects can be represented in AutoCAD such as 3D faces, edge surfaces, ruled surfaces, tabulated surfaces and surfaces of revolution. One object may comprise a number of different surfaces connected together. The following sections will describe each of these surfaces, paying particular attention to how these surfaces are defined and constructed, as well as the parameters required to construct them.

8.5.1 3D face

3D faces are flat surfaces created and bound by three or four edges, principally implying triangular or rectangular shapes. The AutoCAD command, 3DFACE is used to insert surfaces by defining three or four points and the surface will be created bound by those three or four edges. The default and required parameters for this command are first point, second point, third point and fourth point, then the 3D face flat surface is created. In the case where the surface is bound by three edges, upon selection of the third point,

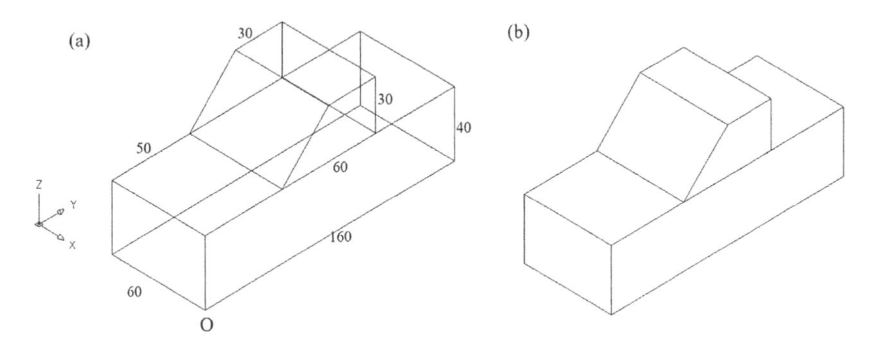

Figure 8.5 (a) Wireframe model (b) Same model with surfaces and hidden lines removed.

the command is terminated by pressing Enter (↵) and the 3D face flat surface is created within the three edges. The 3D face can be best illustrated by retrieving the sample 3D wireframe model that was created in the previous section and then inserting the 3D faces on all the edges that were created, as long as they are bound by three or four edges. Having inserted the 3D faces, this can be proved by invoking the HIDE command and the result will be displayed in Figure 8.5. To unhide, the REGEN command can be used to bring back all the edges that were previously defined. The wireframe model can be reproduced on the same page and then 3D faces added to the one on the right and then after applying the HIDE command, the 3D wireframe model on the left will remain the same but the one on the right with 3D faces will conceal all hidden edges. This is probably the most significant difference between wireframe and surface models.

8.5.2 Edge surface

Edge surfaces are almost similar to the 3D face in that they are bound by four edges except that edge surfaces are created by four adjoining and connected lines or arcs. It is important and necessary to ensure that the edges are properly connected at their endpoints otherwise any opening may result in no surface or improper surface created. The AutoCAD command to create such surfaces is EDGESURF followed by selecting the adjoining arcs and lines, from which the surface will be created. Unlike 3D faces, edge surfaces are formed by a series of lines forming a mesh, the density which is controlled by the two variables, **Surftab1** (horizontal surface mesh density) and **Surftab2** (vertical surface mesh density), equivalent to the aspect ratio of the mesh. The default value for both *surftab* values is 6, implying that there will be six columns and six rows of the mesh across the surface created. The higher the values, the more precise and smoother the surface. Figure 8.6 shows an illustration of the edge surface created from a series of arcs connected to a series of lines. Once the adjoining arcs and lines are selected, the

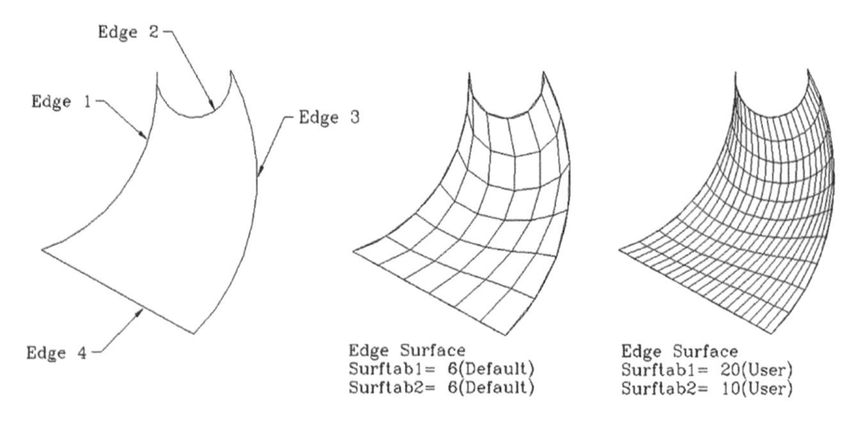

Figure 8.6 Edge surface created from adjoining lines and arcs.

edge surface is created. The user can also experiment with different *surftab* values to see the effect of the resulting surface as shown in Figure 8.6. The *surftab* values set at any particular point will be applicable to that current session of drawing only.

8.5.3 Ruled surface

A ruled surface is almost equivalent to an edge surface except that the edges or lines required to generate the surface do not have to be adjoining. The ruled surface is used to create a mesh that represents the surface between two lines or curves. The AutoCAD command to generate this kind of surface is RULESURF followed by selecting the first defining curve and the send defining curve, and then the mesh is created. The appearance of the mesh is dependent on where the first and second defining curves are selected and this is demonstrated in Figure 8.7.

The edges that are required in this method can be lines, arcs, splines, circles, or polylines. If one of the edges is closed, then the other edge must also

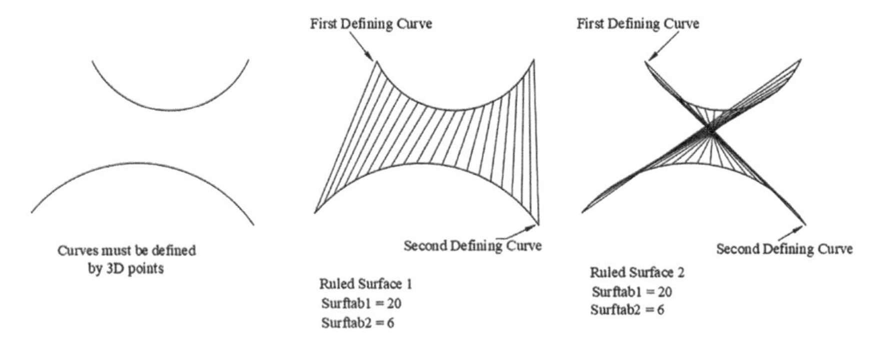

Figure 8.7 Illustration of the ruled surface.

be closed. Users can also use a point as one edge for either an open or a closed curve. The MESHTYPE system variable sets which type of mesh is created. Mesh objects are created by default while setting the variable to 0 creates a legacy poly-face or polygon mesh. For closed curves, the selection does not matter. If the curve is a circle, the ruled mesh begins at the 0-degree quadrant point, as determined by the current x-axis plus the current value of the SNAPANG system variable, that is used to set the snap and grid rotation angle for the current viewport relative to the current UCS. For closed polylines, the ruled mesh starts at the last vertex and proceeds backwards along the segments of the polyline. Creating a ruled mesh between a circle and a closed polyline can be confusing. Substituting a closed semicircular polyline for the circle might be preferable.

The ruled surface mesh is constructed as a 2 by n polygon mesh. RULESURF places half the mesh vertices at equal intervals along one defining curve, and the other half at equal intervals along the other curve. The number of intervals is also specified by the **Surftab1** system variable as in the case of the edge surface. It is the same for each curve; therefore, the distance between the vertices along the two curves differs if the curves are of different lengths.

The n direction of the mesh is along the boundary curves. If both boundaries are closed, or if one is closed and the other is a point, the resulting polygon mesh is closed in the n direction and n will be equivalent to **Surftab1**. If both boundaries are open, the n equals Surftab1+1, because the division of a curve into n parts requires $n + 1$ tabulations. The (0,0) vertex of the mesh is the endpoint of the first selected curve nearest the point used to select that curve. Selecting objects at the same ends creates a polygon mesh. It should be noted that the ruled surface is a patch that is created between two 3D entities and that the orientation of the patch depends on the sequence of selection of the defining curves.

8.5.4 Tabulated surface

A tabulated surface is created from extruding a continuous polyline in a given direction/orientation by generating a mesh that is swept along a straight path. This is done by selecting a line, arc, circle, ellipse, or polyline to sweep in a straight path followed by selecting a line or polyline to determine the first and last points of a vector that indicates the direction and length of the polygon mesh. Similar to the ruled surface, the MESHTYPE system variable sets which type of mesh is created. The AutoCAD command to execute this command is TABSURF, followed by selecting the path curve (continuous polyline) and a direction vector as shown in Figure 8.8. The surftab1 system variable is handled in the same manner as the previous cases. The path curve defines the approximated surface of the polygon mesh, which can be a line, arc, circle, ellipse, or 2D or 3D polyline. The mesh is constructed starting at the point on the path curve closest to the selection point. The direction vector specifies a line or open polyline that defines the

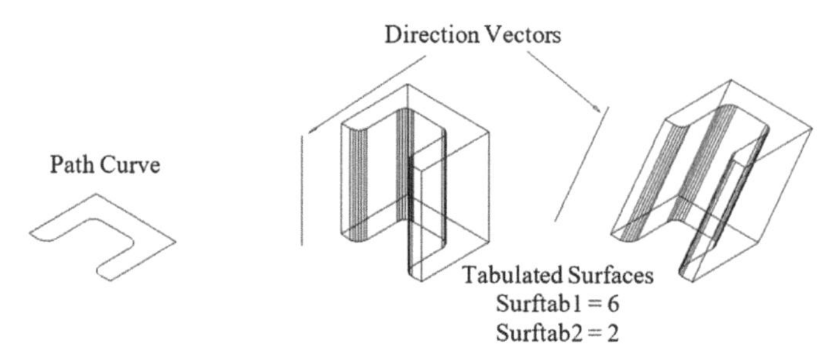

Figure 8.8 Tabulated surface (extrusion) from continuous polyline (path curve).

direction of the sweep. Only the first and last points on a polyline are considered, and intermediate vertices are ignored. The direction vector indicates the direction and length of the shape to be extruded. The end selected on the polyline or line determines the direction of the extrusion. The original path curve is constructed with wide lines to help users visualize how the direction vector dictates the construction of a tabulated mesh.

8.5.5 Elevation

Elevation is another useful method in surface modelling almost similar to tabulated surfaces in that it sets the elevation and thickness of new objects. It creates surfaces that are extruded in the (z) direction. The AutoCAD command for generating extruded surfaces is ELEV in which the elevation is the default (z) value above (+) or below (–) the XY plane while the thickness is the height of the extrusion. Objects drawn after setting the two variables in the particular session of AutoCAD are the only ones affected and will be constructed at the elevated figure. This value is stored in the ELEVATION system variable. To use this facility more efficiently, it is recommended that users leave the elevation set to zero and control the XY plane of the current UCS with the UCS command instead, otherwise it will distort future objects if the user forgets to revert to the original elevation.

As will be seen in later chapters on managing and organizing CAD drawings, the elevation setting is the same for all viewports regardless of their UCS definitions. New objects are created at the specified (z) value relative to the current UCS in the viewport. The elevation is particularly useful in architectural models where window openings or lintels should be drawn elevated from the ground. Figure 8.9 is an illustration of objects extruded in the (z) direction and also a demonstration that text cannot be extruded.

8.5.6 Surface of revolution

A multitude of objects in real life is symmetrical about a given axis. Depending on what the objects are, they can either be modelled using the surface of

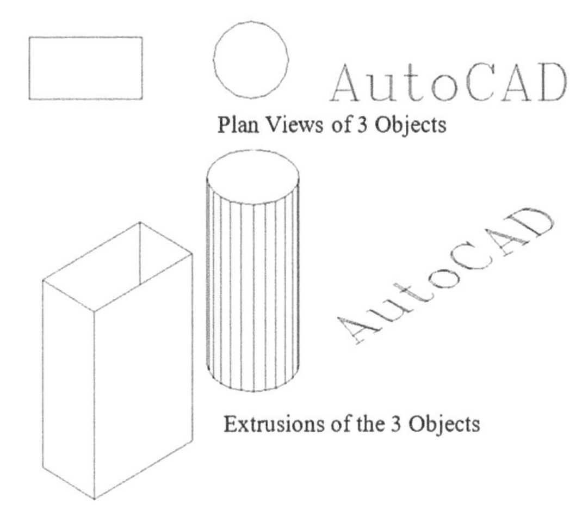

Figure 8.9 Objects extruded in the Z-direction.

revolution (hollow objects) or solid of revolution (solid objects, as will be seen in the next chapter). The AutoCAD command for generating a surface of revolution is REVSURF, which requires six parameters to be defined before it can be revolved. The object to be revolved is the path curve that is generated from one of the elevations, either front or plan and needs only be one-half of the view. This path curve is generally created from lines and arcs which need to be converted to a single polyline (continuous path curve). This can be done by first creating the required lines and arcs, then using the PEDIT command, the lines and arcs can be converted to polylines and then joined to become one, using the same command as covered in Chapter 3. As discussed earlier, the Surftab1 and Surftab2 system variables also need to be set for the surface of revolution. The default values for both are 6 but this may result in a hexagonal-shaped object instead of a round object as is expected from a revolution. To achieve a smooth rounded object, the two system variables need to be set to much higher values.

The other important parameter that needs to be defined is the axis of revolution, which can be a construction line through the mid-section of the object (splitting half of the path curve). AutoCAD also includes two additional parameters, the start and end angles. This is meant to construct surfaces of revolution that are not necessarily fully circular, such as a hollow shaft that is sliced open for analysis. Otherwise, the default start and end angles for the REVSURF command are 0°and 360° (full circle), to revolve the complete object. Figure 8.10 shows a surface of revolution for a stepped shaft, starting with the constructed path curve, comprising lines, fillets and chamfers as well as a construction centre line representing the axis of revolution and finally the resultant surface of revolution obtained by tilting the object to a suitable viewing point using WCS or using the preset 3D views.

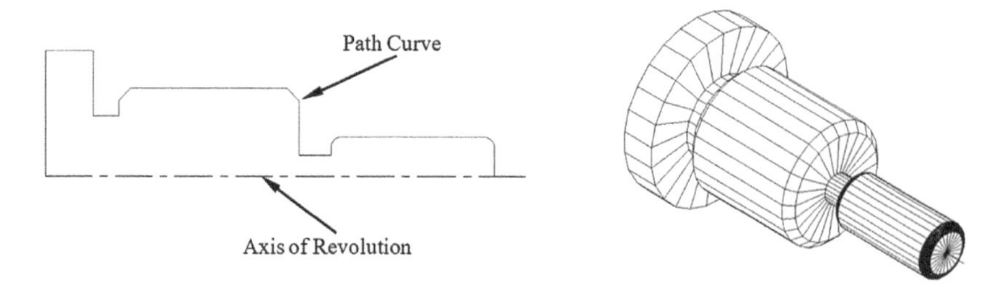

Figure 8.10 Surface of revolution for a stepped shaft.

8.6 APPLICATIONS AND LIMITATIONS OF WIREFRAME AND SURFACE MODELS

Although there are some limitations as discussed in the previous sections of this chapter, wireframe and surface models find wide applications in various disciplines of industry, from designing to production and manufacturing to the construction industry. The phrase 'wire frame' is derived from designers using metal wire to represent the 3D shapes of solid objects. 3D wireframe computer models allow for the construction and manipulation of solids and solid surfaces by depicting the shape or *skeleton* of a real-life object. Using a wireframe model allows for the visualization of the underlying design structure of a 3D model, albeit in *skeleton* form.

Since wireframe models are relatively simple and fast to generate, compared to surface and solid models, they are often used in cases where a relatively high screen frame rate is needed (for instance, when working with a particularly complex 3D model, or in real-time systems that model exterior phenomena). Wireframe models are used as a foundation for both surface and solid models as seen and applied in the case of 3D faces described earlier. When more graphical details are required, surface textures can be added automatically after the completion of the initial construction of the wireframe to enable the designer to quickly analyse objects without the cumbersome requirements for surfaces and solids.

Wireframe models are also well-suited and widely used in programming tool paths for direct numerical control (DNC) machine tools in CAM. Wireframe models have been used extensively in video games to represent 3D objects during the 1980s and 1990s, when filled 3D objects (solid) would have been too complex to derive and model with the computers at the time (Simmons et al., 2020). Since the turn of the new millennium, wireframe models have found other applications such as in the transport sector, both automobile and aerospace, in modelling motor vehicles and airplanes. They are also useful as input to finite element modelling to determine the stresses and strains of objects under loading and the prediction of performance using simulations.

Even in architectural designs, models of buildings are now commonly developed using wireframe models. Although wireframe models are as straightforward to create as 2D drafting, the major challenge for this type of modelling is that it does not exhibit what the real-life object is bound by, nor does it show the contents of the real, hence the need for surface and solid models. In addition, depending on the model under consideration, the multitude of lines forming the wireframe model also creates confusion because these cannot be hidden in the absence of surfaces to conceal them in order to improve visualization.

In view of the limitations of wireframe models, surface models are a better representation of real-life objects in that they contain information on the surfaces and provide better visualization of the object, especially with hidden lines removed. Surface modelling enables users to visualize real-life objects at different angles with the option of removing edges behind surfaces to improve vision and analysis. Just like wireframe models, surface models have wide and varied applications in architectural designs, consumer products, marine vehicles, body panels of automobiles and aircraft structures. Surface modelling is a relatively more complex method for modelling 3D objects compared to the wireframe in that various parameters must be considered and appropriate surface patches selected.

Despite such complexities, surface models have much less ambiguous display functionalities compared to wireframe models, but are not as much or sophisticated as solid models. The surface modelling technique often involves conversions between various 3D modelling types and the different surface options to carefully select before a complete surface model can be created. Generally, these processes involve the generation of a model combining the 3D surfaces and solids, conversion of the model to procedural surfaces, taking advantage of associative modelling, validation of imperfections with surface analysis tools and confirmation of appropriate surface patches selected as well as rebuilding surfaces of objects to apply smoothness to the objects by appropriately adjusting the *surftab* system variables.

For dealing with curves, surface modelling techniques make use of B-splines and Bezier curves mathematical techniques. One of the limitations of surface models is that they cannot be sliced open for analysis like solid models. In addition, they are also limited to depicting the shape and bounding surfaces but not the contents or materials that form the object, hence the need for solid modelling. The objects used in surface modelling can be geometrically incorrect, unlike in solid modelling, where it needs to be correct and based on the solid primitives as will be seen in the next chapter. Apart from architectural illustrations, surface modelling is also used in 3D animation, particularly in games. Figure 8.11 shows some of the applications for wireframe and surface models.

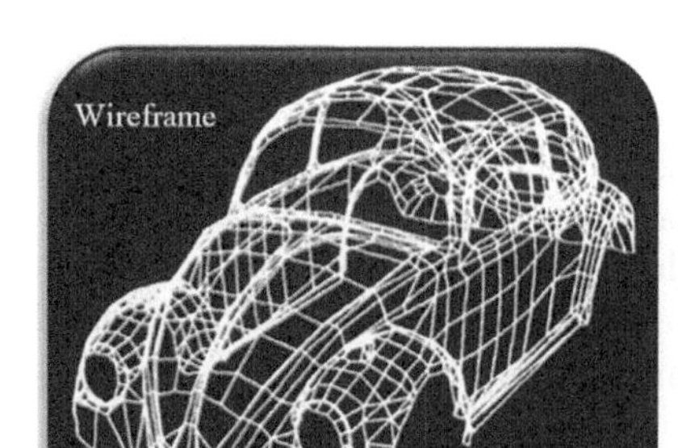

Figure 8.11 Applications for wireframe and surface models.

8.7 SUMMARY

Three-dimensional models are useful tools in CAD to better visualize and understand how a real-life object can be represented, particularly to laymen in engineering and technical drawings. Most decision-makers in companies are of management and accounting backgrounds, hence the need to represent objects in a manner they understand in order to facilitate decision-making in businesses. This chapter was an introduction to 3D modelling, outlining how the UCS and WCS coordinates can be manipulated to view objects at different angles and viewing points, or by alternatively using the preset 3D views, SE, SW, NW and NE isometric orientations. This was followed by the outline of the basic representations of objects, that is, wireframe and surface modelling, how they are created and used as well as some of the common applications for the two, which include architectural modelling, aerospace and automobile industry as well as consumer products in manufacturing.

While wireframe models are generally easier to create, they are limited in that they only show the skeletal structure of an object and the multitude of edges (wires) can be quite confusing to the user as those behind others cannot be hidden. Surface models are a little more involving to create as there are several options or surface patches that the user needs to correctly pick and use. However, they are better in terms of visualization as edges behind surfaces can be concealed to improve vision. Unfortunately, they are also limited to shapes, surfaces and structure of the object but not their contents, hence the need for solid models, which will be handled in the next chapter.

8.8 REVIEW EXERCISES

1. In the earlier versions of AutoCAD, users needed to go through the cumbersome process of specifying viewing coordinates using the VPOINT command and then work out a suitable (x, y, z) coordinate

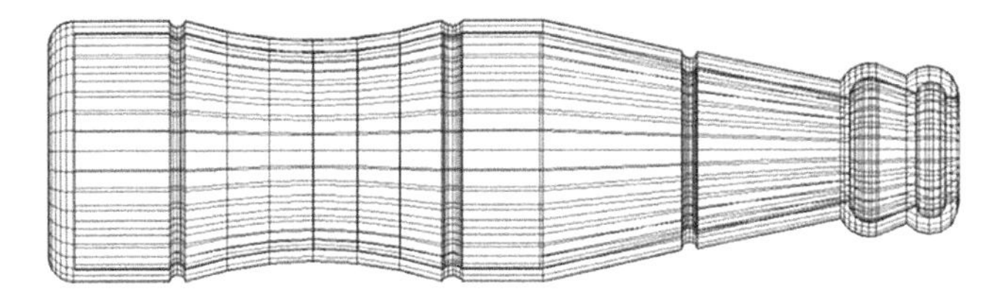

Figure 8.12 Plastic bottle created using the surface of revolution.

to get the desired orientation of a 3D object. The newer versions have simplified this to the use of the WCS or the preset 3D views, SE, SW, NW and NE. Explain how this works and how a suitable orientation of a 3D object can be achieved for maximum visualization.

2. Wireframe and surface models are the basic representation of objects. Outline the differences between the two, highlighting the limitations of each in modelling real-life objects.
3. Outline the six most commonly used techniques in surface modelling giving examples of their practical applications in modelling objects in engineering.
4. List and explain at least two everyday practical applications for wireframe and surface models.
5. List and explain the six parameters required to define a surface of revolution.
6. Figure 8.12 shows a plastic bottle created using the surface of revolution. Using the parameters listed in (5), generate the surface model of the bottle using AutoCAD, explaining each step and why the particular parameters were chosen to model the bottle.

Chapter 9

Three-dimensional solid modelling

9.1 INTRODUCTION

There are several ways to improve the visualization of objects in three dimensions, the most commonly used techniques being wireframe, surface and solid modelling. Chapter 8 focused on wireframe and surface models, in which it was highlighted that, although these two techniques are quick to generate and view three-dimensional objects, they are not only limited to displaying the shapes, structure and surfaces of objects but sometimes errors occur resulting in incorrect representations. The ultimate representation of objects is solid modelling in which, not only the shapes, structure and surfaces are represented but the internal contents as well. Specifying materials of objects ultimately enables users to analyse them in terms of mass and weight, mass moments of inertia, interference checks on parts fitting together and the ability to slice open and view the internal contents, especially in objects that are assembled and comprising several components such as typical machines. The aim of Solid Modelling is to make sure that all surfaces modelled are geometrically correct. Solid modelling is considered the most complex aspect of 3D CAD because it requires the CAD software to simulate the object to be representative of real-life objects (Nyemba, 2000).

This chapter focuses on 3D solid modelling paying particular attention to and focusing on the two most commonly used techniques of boundary representation (B-Rep) based on the boundaries of an object and constructive solid geometry (CSG) based on Boolean algebra and its applications in combining, subtracting or intersecting solid primitives to form solid objects, both in terms of single components or an assembly of several components. In addition, the practical applications are also outlined, leading to a case study for the failure analysis of a connecting rod as well as another worked-out example in the form of a tutorial in Chapter 10. Although the use of 3D solid models is limited, according to studies and research carried out, it is a further demonstration of the need to not only invest in CAD systems but also to fully utilize all the facilities available in order to benefit from the complete suites of CAD systems (Nyemba, 2000).

DOI: 10.1201/9781003288626-9

Being the ultimate representation of objects, solid models are involved with the construction, modification and analysis of solid primitive shapes that can be extruded in three dimensions, revolved, combined, intersected or subtracted to form composite solid objects. Models of assembly such as machines can also be constructed in 3D while generated 3D models can be used to automatically generate 2D orthographic views. The ability to open up by slicing models of objects or assemblies, allows the analysis of particular sections or regions of the model for interference, especially for components that fit into each other. Ultimately, the definition or addition of materials and rendering to different forms provides a realistic display of the models under development.

Until the 1990s, AutoCAD was largely used for 2D drafting, with very little or no 3D modelling. In the early 1990s, when AutoCAD Release 11 and 12 were developed, they were provided with an add-on facility, AutoCAD Modelling Extension (AME) for 3D modelling. From 1994, when AutoCAD Release 13 was unveiled, AME became an integral part of the software till this AutoCAD 2021 and is expected to remain so in future (Kennedy, 2014).

9.2 BOUNDARY REPRESENTATION (B-REP) AND SOLIDS OF EXTRUSION

In 3D solid modelling and CAD in general, boundary representation (B-Rep) is a technique for representing shapes using their limits or boundaries. A solid is generally represented as a collection of connected surface elements which define the boundary between its interior and exterior points. A boundary representation of a model therefore comprises topological components, which can be faces, edges or vertices and the connections between them, along with their geometric definitions such as surfaces, curves and points, respectively. A face is a bounded portion of a surface while an edge is a bounded piece of a curve and a vertex lies at a point. Other elements are the shell, which is made up of a set of connected faces, the loop, which is comprised of a circuit of edges bounding a face and loop-edge links, which are sometimes referred to as winged edge links or half-edges which are used to construct the edge circuits (Krysl & Ortiz, 2001).

The basic philosophy and technique for boundary representation can be traced to the 1970s when a CAD expert from Cambridge and computer visionary from Stanford developed the concept which later diversified to BUILD, a solid modelling system, on which many such systems are based and the commercial systems ROMULUS, the forerunner of Parasolid and ACIS (almost similar to AME). Parasolid and ACIS are the basis for many of today's commercial CAD systems (Lee et al., 2005). The technique was refined to enable them to work with hybrid models wire-frames, sheet objects and volumetric models during the early 1980s. Numerous other solid modelling systems, involving boundary representation, were later developed and to date, these continue to be developed, depending on customized requirements throughout the world. The development of reliable commercial B-rep kernel systems like Parasolid and ACIS as well as Open CASCADE and C3D that were later developed has led to widespread adoption of B-rep for CAD (Zhuo et al., 2012).

Boundary representation is thus, basically a local representation connecting faces, edges and vertices. An extension of this was to group sub-elements of the shape into logical units called geometric features, or simply features, on which many other developments, allowing high-level 'geometric reasoning' about shape for comparison, process-planning, manufacturing, etc. are based (Zhuo et al., 2012). Boundary representation has also been extended to allow special, non-solid model types, referred to as non-manifold models. Normal solids found in nature have the property that, at every point on the boundary, a small enough sphere around the point is divided into two pieces, one inside and one outside the object (Stroud, 2006). However, non-manifold models break this rule in that, an important sub-class of non-manifold models are sheet objects which are used to represent thin-plate objects and integrate surface modelling into a solid modelling environment (Stroud, 2006).

B-Rep models are based on the boundaries of objects which can be extruded in the same manner as the extrusions discussed in Chapter 8. The key aspect is to properly define the path curve (boundary of the object). A typical example of this is in architectural design, where walls can be extruded from a path curve to a suitable height. Figure 9.1 shows three

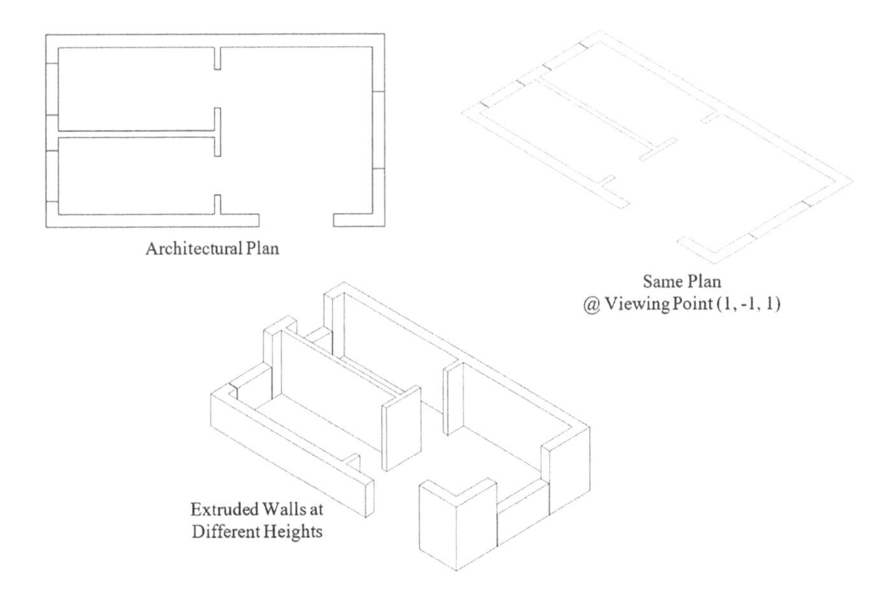

Figure 9.1 Illustration of boundary representation in architecture.

displays. The first display is an example plan of a building showing walls, door and window openings. For practice, this can be created in the same manner as explained in earlier chapters by using either line or polyline, but whichever is used, the segments that need to be extruded together to the same height must be *joined* to form one continuous path curve using the PEDIT command as explained previously. The second display is the same architectural plan but at a suitable viewing point. At this stage, the AutoCAD command, EXTRUDE can be used and then the user must select those path curves with the same height and specify the height and direction of extrusion to obtain the resulting third display (with hidden lines removed). Specifying the heights differently helps to manage door and window openings. If needed, to construct from an elevated window opening, then ELEV must be set to a suitable value.

9.3 CONSTRUCTIVE SOLID GEOMETRY (CSG)

Constructive solid geometry (CSG), which was previously referred to as computational binary solid geometry is a technique used in solid modelling to create complex objects or models by using Boolean operators to combine simpler objects, commonly referred to as solid primitives. This potentially generates and results in visually complex objects by combining a few primitive ones. In 3D computer graphics and CAD, CSG is often used in procedural modelling. CSG can also be performed on polygonal meshes, and may or may not be procedural and/or parametric (Zuo et al., 2014). Unlike B-Rep models which are confined to generating solid objects based on their boundaries, CSG can be used to create virtually any solid model as objects are believed to be made up of simple shapes that can be combined in different ways. However, CSG may be more complex and time-consuming than B-Rep.

9.3.1 Solid primitives and building blocks

The simplest solid objects or geometric shapes used in CSG are the building blocks or geometric primitives. These range from the simplest such as the box or prism followed by cylinder spheres and cones to the more complex ones such as the torus and wedge (Nyemba, 2000). The range of solid primitives available gives the modeller a wide variety of choices from which the most appropriate and optimum number of primitives should be used, the fewer there are, the more efficient and faster the development of the model. However, the set of allowable primitives is limited by each CAD software package. Some software packages allow CSG on curved objects while other packages do not (Zuo et al., 2014). In earlier versions of AutoCAD up to Release 12, such primitives were accessible through the advanced modelling extension (AME) add-on facility and at that time, each

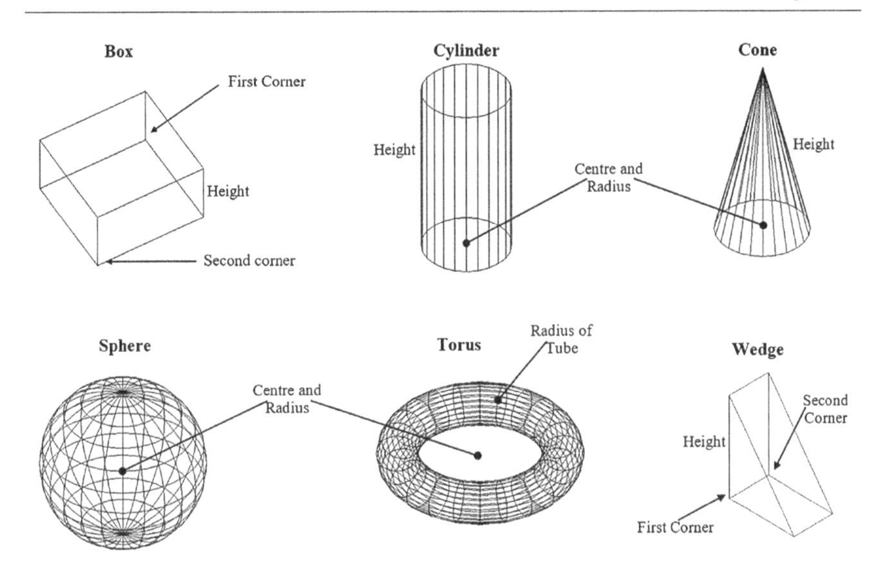

Figure 9.2 Solid primitives showing required parameters.

of them was prefixed with SOL, for example, SOLBOX, SOLCYLINDER, SOLWEDGE etc. and these prefixed words were the commands used to create them.

In the later versions of AutoCAD in which AME is an integral part of the package, the solid primitives can be accessed by simply typing the name as a command, such as BOX, CYLINDER, WEDGE etc. or more simply from the **Modelling** icon menu. Ordinarily and by default, this menu is not loaded when AutoCAD starts but can be turned on in the same manner as other icon menus, that is, RMB on the empty space adjacent to the slide menus, then select AutoCAD and tick Modelling. Similarly, the **Solid Editing** icon menu which contains Boolean operations (union, intersection and subtraction) can also be ticked and positioned appropriately for use. What is critical in the use of such primitives is to ensure that the right size is defined and that the primitive is placed in an appropriate position to execute the Boolean operations otherwise an incorrect model may arise, hence the knowledge and experience to use as minimal as possible, the number of primitives that make up an object. Each primitive can be loaded using defined parameters such as those shown in Figure 9.2.

9.3.2 Boolean algebra and operations on solid primitives

The combination of the various primitives in Figure 9.2 is based on Boolean algebra. It is assumed that CAD users at this stage understand the philosophy behind this branch of algebra or mathematical logic in which the values of the variables are the truth values true and false, usually denoted 1 and 0, respectively. Instead of elementary algebra, where the values of the variables are numbers and the prime operations are addition and multiplication, the key operations in Boolean algebra are the conjunction (and), the disjunction (or) and the negation (not). It is therefore a formal way of describing logical operations, in the same way that elementary algebra describes numerical operations. As a recap, Boolean algebra owes its origins to George Boole in Mathematical Analysis of Logic (Boole, 2011). Boolean algebra has been the foundation in the development of digital electronics and is provided for in all modern programming languages. It is also used in probability and statistics as well as set theory, the basis on which CSG was founded and used in this book.

The three basic principles of operation on which CSG is based are: Union (addition), intersection and subtraction. In CSG, objects are thus first positioned in appropriate places and then united, intersected or subtracted to get a resulting solid model. This is also the same basis on which mathematics of sets or geometric transformations of those sets are carried out. The three Boolean operations are simplified and represented mathematically as illustrated in Figure 9.3, based on set theory and mathematics of sets. Typically, a primitive can be described by a procedure which accepts some number of parameters as shown in Figure 9.2, for example, for a box, the first and second corners as well as the height of the box; for a torus, the centre and radius of the torus as well as radius of the tube; for a cylinder, the centre of the base, radius and height.

Instead of the illustrations of the Boolean operations on circles as in Figure 9.3, the three operations can be carried out on appropriately positioned solid primitives to obtain the required composite solid models as illustrated in Figure 9.4. The AutoCAD commands for the three are UNION, INTERSECT and SUBTRACT or alternatively selecting these from the **Solid Editing** icons, as described earlier. In so doing, it is important to follow the prompts on the Command prompt to ensure that the correct operation is being executed. For example, in the case of union, the system requires the user to select the primitives to be united, one after the other, whereas, for subtraction, the primitive to subtract from is selected first followed by the one to subtract. The Command prompt provides adequate guidelines for this.

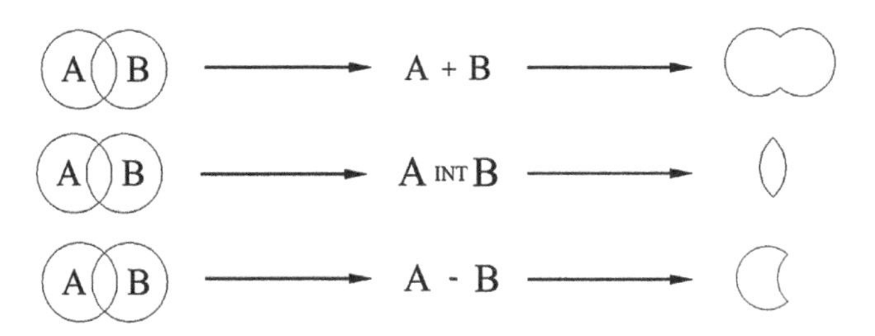

Figure 9.3 Boolean operations of union, intersection and subtraction.

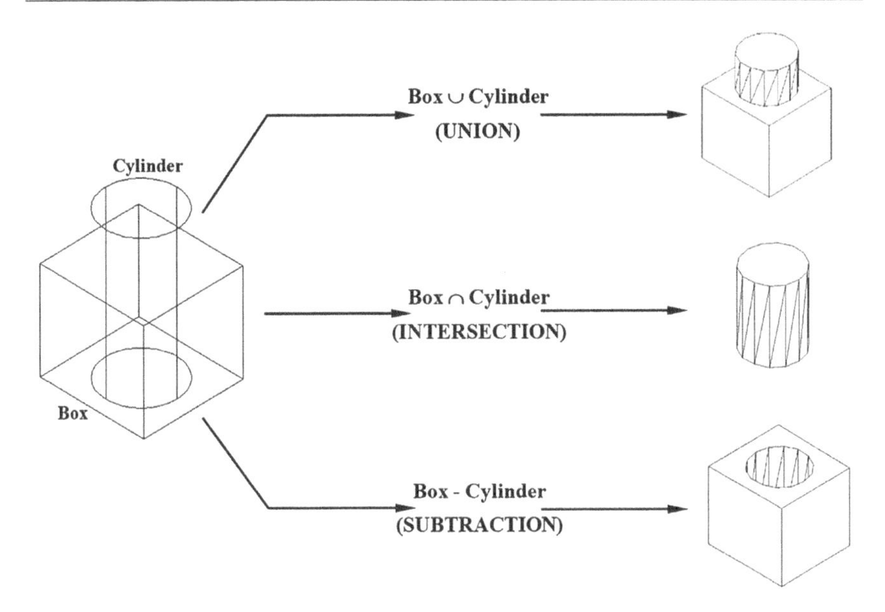

Figure 9.4 Boolean operations on solid primitives (box and cylinder).

9.4 SOLID OF REVOLUTION

Similar to hollow and symmetrical objects as covered in surface modelling in the previous chapter, many solid components are also symmetrical about a given axis. The method and sequence of generating a solid of revolution are similar to that for the surface of revolution where a path curve must be defined followed by the axis of revolution, start and end angles. In later versions of AutoCAD, the user can actually drag the revolution from the desired start to end angles, otherwise for a normal full solid, the default option for a full circle (360°) is chosen.

Figure 9.5 shows a picture of a solid flange taken from different angles. To model this solid flange, the first thing would be to create the path curve consisting of a continuous polyline and the axis of revolution as shown in the orthographic elevation in Figure 9.6. The AutoCAD, REVOLVE is then used, followed by selecting the path curve and axis as well as a full circle to get the solid revolved. A suitable view angle can be chosen, such as SE or NW.

Instead of incorporating them on the path curve, fillets and chamfers can be created on the revolved solid model using the 3D commands FILLET and CHAMFER commands. The earlier versions of AutoCAD required the user to set the solid wire density (SOLWDENS), equivalent to the system variables, Surftab1 and Surftab2 in surface modelling but this is now a standard in later versions since the user has the opportunity to visualize the model in different styles such as rendered, realistic, conceptual, hidden etc. as will be discussed in the next sections. In the next step, the holes are added to the

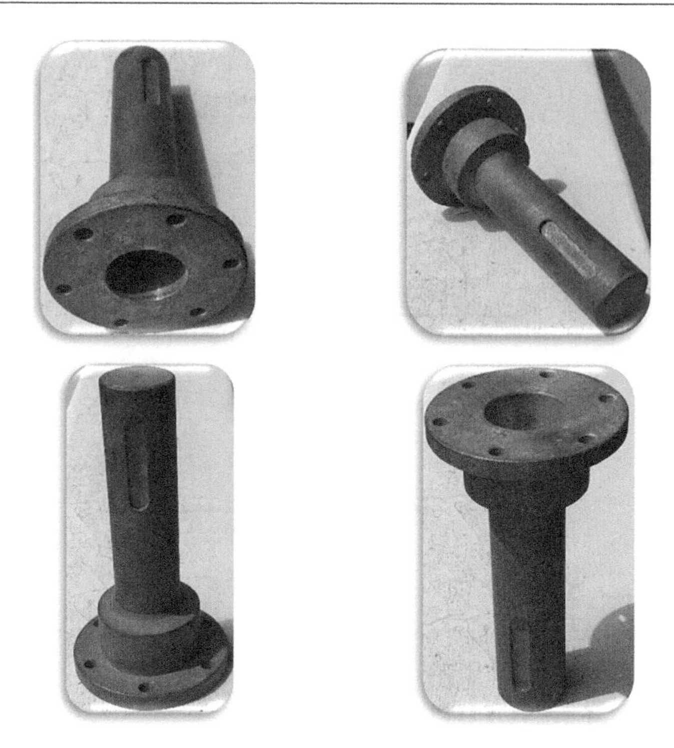

Figure 9.5 Pictures of a solid flange from different angles.

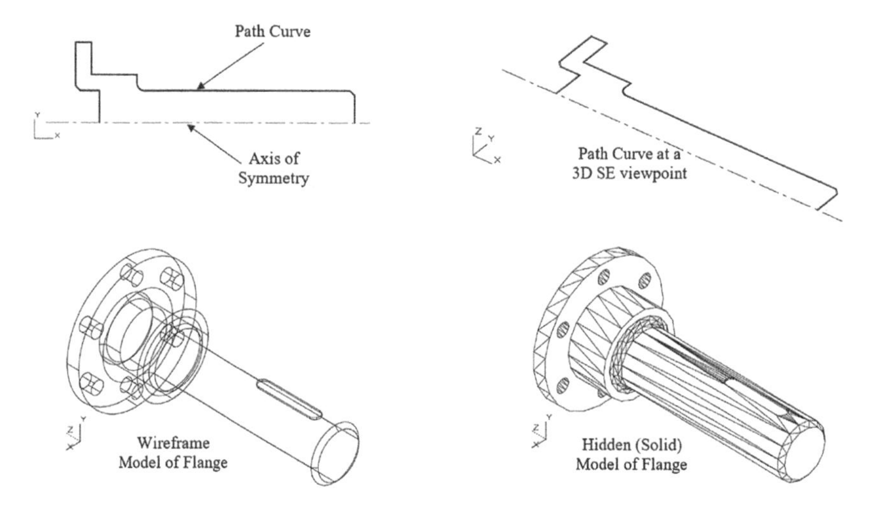

Figure 9.6 Path curve and models of the flange.

base and the keyway on the shaft. The model can be positioned using the WCS or the preset 3D views and then a construction circle that passes through the centres of the holes is drawn followed by creating one of the required cylinders, centred on the construction circle. The rest of the cylinders can be generated by using a polar array, centred at the axis of symmetry and specifying six of them, which will then automatically be generated, followed by subtracting them to form the holes.

However, it should be noted that, since the six cylinders were created using an array, they need to be exploded first before they can be subtracted, otherwise an error is prompted if the user tries to subtract all six cylinders as a composite. The keyway is created in a similar fashion. It consists of a solid box with two solid cylinders at the ends. These three primitives can be created separately from the flange by uniting the two cylinders and the box using SOLUNION (AutoCAD Release 12 and before) or UNION (AutoCAD Release 13 and later). Before subtracting the result of this union, it must be moved from where it was constructed and inserted appropriately on the flange shaft. Construction lines may be necessary to ensure that the resulting union of the cylinders and box are within a position on the shaft that will enable creation of the keyway by subtracting it from the flange model. Figure 9.6 shows the orthographic elevation of the flange path curve and axis of symmetry before revolving it together with the wireframe and hidden models of the solid of revolution of the flange. These can be visualized better using the different styles now readily available in the latest versions of AutoCAD, as described in Section 9.5.

9.5 VISUALIZATION OF 3D MODELS

CAD systems are increasingly becoming more and more useful and popular mainly because of the wide range of facilities that developers continue to add, in particular, the variety of ways in which 3D objects can be viewed. AutoCAD in particular offers various options to improve the visualization of models that include, edges, lighting, shading and rendering. The various visual styles are available under the **View** pull-down menu, followed by Visual **Styles**. Apart from the 3D wireframe and hidden models discussed in the previous section, the other visual styles include **Realistic, Conceptual, Shaded, Sketchy** etc. Rendering is also another option but will be covered together with analysis in the next section.

The differences between these visual styles and their meanings are as follows:

2D Wireframe: Allows for the display of objects using lines and curves in a skeletal form. This visual style is optimized for a 2D drawing environment with high fidelity. However, due to the fact that all lines and edges will be visible including hidden ones, it can be difficult to appreciate the complete structure especially if it is composed of many edges.

Conceptual: Displays 3D objects using smooth shading and the Gooch face style, which allows for transitions between cool and warm colours, rather than dark and light (Al-Rousan et al., 2018). Although the effect is less realistic, the details of the model are easier to see.

Realistic: Displays 3D models using smooth shading and materials.

Shaded: Displays 3D models using smooth shading.

Shaded with Edges: Displays 3D models using smooth shading and visible edges.

Shades of Gray: Displays 3D models using smooth shading and shades of gray.

Sketchy: Displays 2D and 3D models with a hand-sketched effect by using the Line Extensions and Jitter edge modifiers (Mell & Monroy, 2018).

Wireframe: Displays 3D models using lines and curves only as with 2D wireframes but the draw order settings and fills from 2D solid objects are not displayed. In this mode, the view is also not regenerated when the view direction is changed as is the case with the 2D wireframe. It therefore saves time, especially with large 3D models.

X-ray: Displays 3D objects with partial transparency.

The wireframe, realistic and shaded styles provide enhanced 3D performance for panning, zooming and orbiting operations. The VSFACESTYLE system variable controls how faces, solid-fill hatches and gradients are displayed. This consists of three values: 0 – No style applied, 1 – Real: as close as possible to how the model would appear in the real world and 3 – Gooch style of cool and warm colours to enhance the display of faces that may be in shadow and difficult to see in a realistic display. Figure 9.7 shows four of the most commonly used visual styles in AutoCAD, as applied to the solid flange that was developed in the previous section.

9.6 RENDERING AND SPECIFYING MATERIALS FOR 3D MODELS

Apart from the ability to construct and visualize objects in 3D systems including wireframe and surface modelling, the ultimate representation of objects in 3D solid modelling includes the ability to render the objects. Rendering, which is commonly referred to in CAD systems as image synthesis is the technique of developing photorealistic images from 3D models. The **Render** icon menu which includes the options for hiding, projecting lights, adding and editing materials and planar mapping can be activated and loaded onto the AutoCAD screen by right-clicking on the blank space to the top-right of the screen just above the drawing editor, followed by

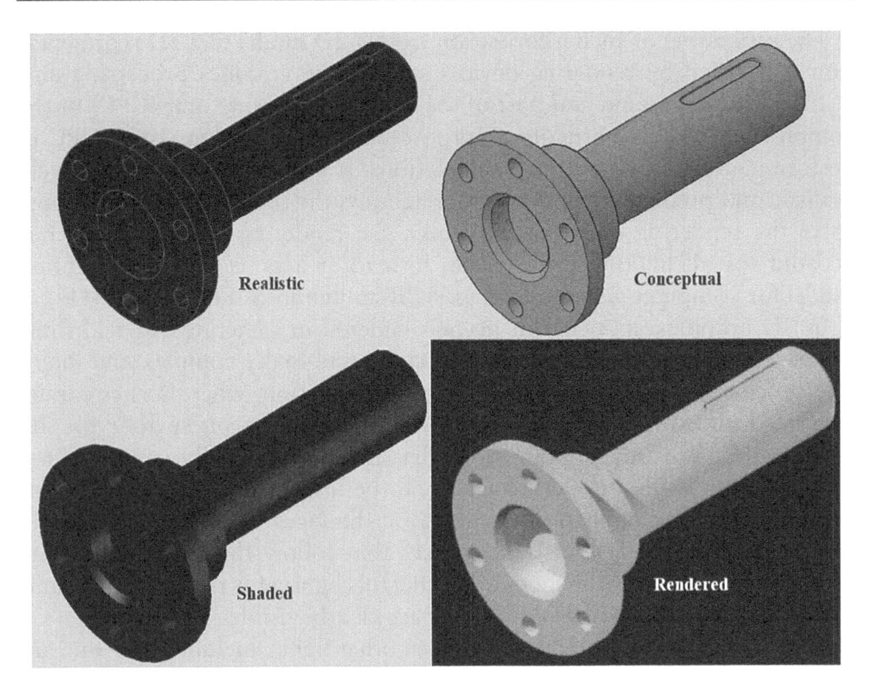

Figure 9.7 Other visual styles for the solid model of the flange.

AutoCAD and ticking the option **Render**. The icon menu can be positioned in a suitable position but it is advisable to only activate it when necessary, otherwise it simply clutters the screen area. Alternatively, all the commands representing the icons can be typed on the command prompt, such as HIDE, LIGHT, MATERIALS etc.

9.6.1 Visual styles and lighting

Multiple models can be defined in a session file containing geometry, viewpoint, texture, lighting and shading information describing the virtual scene. The data contained in the session file are then passed to a rendering program to be processed and output to a digital image or raster graphics image file. Rendering is almost similar or analogous to the idea of an artist's impression of a scene, where it can also be used to describe the process of calculating effects in a video editing program to produce the final video output. Rendering is the last major step in the graphics pipeline, giving models and animation their final appearance. 3D solid modelling and in particular, rendering have been on the increase since the advent of computer graphics (Verma & Walia, 2010). Although the techniques for developing rendered images vary from one CAD system to another, the general challenges encountered in producing a 2D image from a 3D representation stored in a scene file are the same.

The processing of such information from a 2D model to a 3D representation is handled by rendering devices such as the graphics processing unit (GPU), which is an integral part of the central processing unit (CPU) of the computer. A GPU is normally a purpose-built device that assists a CPU in performing complex rendering calculations. If a scene is to look relatively realistic and predictable under virtual lighting, the rendering software must solve the rendering equation. However, the rendering equation does not account for all lighting phenomena. Instead, it acts as a general lighting model for computer-generated imagery (Remondino & El-Hakim, 2006).

In 3D graphics, models can be pre-rendered or generated in real time. However, pre-rendering is a slow and computationally complex and intensive process that is typically used for movie creation, where scenes can be generated ahead of time, while real-time rendering is often done for 3D video games and other applications that must dynamically create scenes. The real-time rendering performance can be improved by using 3D hardware accelerators. In shaded visual styles, the faces of models are usually illuminated by two distant light sources that follow the viewpoint as the user navigates around the model. This default lighting is designed to illuminate all faces in the model so that they are clearly visible.

Default lighting is available only when other lights, including the sun, are turned off. Users can add light points using the point light option on the **Render** icon menu or the LIGHT command and position it appropriately around the model to improve visibility. The **Hidden** and **Sketchy** visual styles, as described earlier, automatically change the colour of solid hatches to the background colour, effectively making them invisible. The original colour can be regenerated to make them visible again by setting the VSFACESTYLE system variable to 1 or 2. Users also have the opportunity to create their own customized visual styles by changing the properties of a particular visual style in the Visual Styles Manager. However, it is recommended to create new visual styles rather than modifying the predefined ones.

When a visual style is applied or its settings changed, the respective viewport is automatically updated to reflect those changes. Any changes made to the current visual style are then saved in the drawing. This will be handled in detail in Chapter 12 on the management of models and drawings for output.

9.6.2 Specifying and editing materials to 3D solids

CAD 3D solid modelling systems allow users to add materials to any 3D solid model, not only to give the model a realistic appearance but also to determine various properties of the model under development such as mass moments of inertia, weight etc. This will be dealt with in more detail in the next chapter which is devoted to a CSG tutorial and worked-out example. In general, AutoCAD allows users to add materials to the entire object but there are also options to add materials to selected faces or individual components, as will be dealt with in the section.

To demonstrate this facility, the flange which was modelled in Section 9.4 as a solid of revolution will be used as an example. The users must ensure that they are in a 3D modelling space and an appropriate viewing point for maximum visibility. The preset viewing points such as SW, SE, NE and NW can be selected from the **View** pull-down menu followed by the **3D Views** and then selecting the appropriate one. The materials browser can be selected from the Render icon menu or by typing the AutoCAD command, MATERIALS on the command prompt. This brings up the materials browser palette with a list of materials available for use in the model under development. A categorized list of materials can be found on the **Home** tab which will show only materials of that category in the palette, for example, selecting Wood will show only wood materials in the palette. This helps in narrowing the search for desired materials such as ceramic, concrete, glass, metal etc. some of which can be further categorized or refined into specific sub-categories as shown in Figure 9.8.

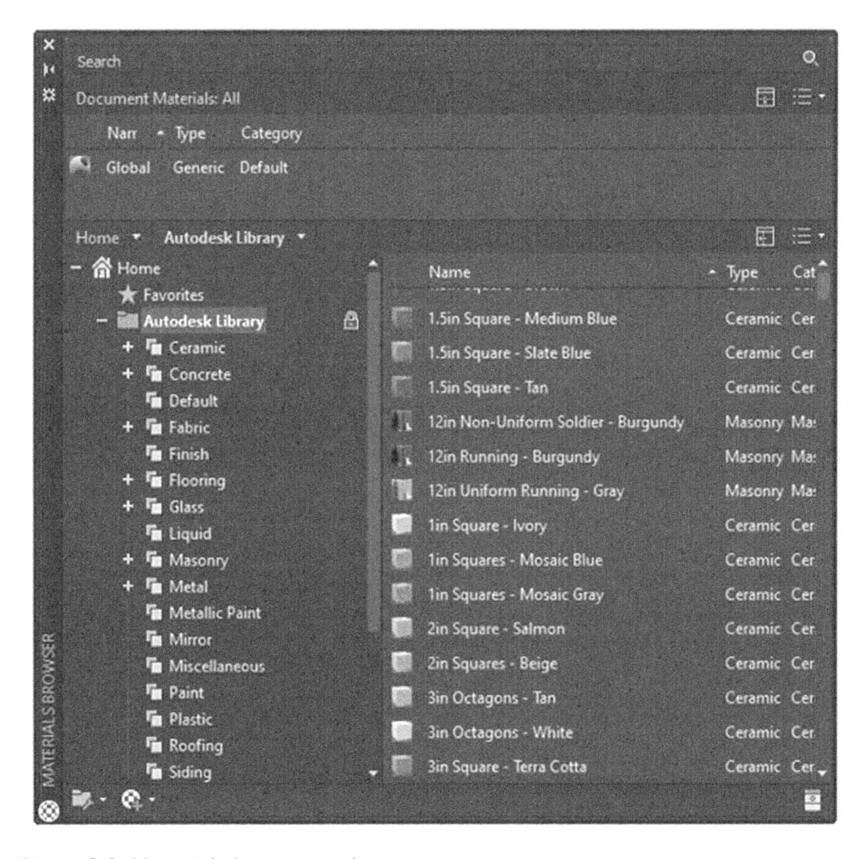

Figure 9.8 Materials browser palette.

The list provided in AutoCAD is quite comprehensive with lots of materials but if necessary, custom materials can be added as well. Once a suitable material has been identified, adding it to the model under development is done by selecting it and then dragging it to the component or model. Alternatively, the model or component can be selected first and then right-clicked on the material to be added from the Materials Browser and then selecting Assign to Selection from the right-click menu. For better visibility and clarity, it is best to set the Visual Style to Realistic and turn on the Material/Texture in the animated image, otherwise the material effect will not be visible on the objects.

The method of drag and drop is limited to applying materials to a whole object but there are instances where the same object may need to have two or more different materials, for example, the inner and outer wall plastering in architecture or different paints on walls, exterior and interior paints are usually different. In this case, the material can be added on individual faces or walls.

The procedure for applying materials on faces is slightly different from that of applying materials on a whole object. In the case where a user wishes to add two different paints on the interior and exterior walls, the paints to be added are selected from the materials browser. The paints can be selected from an appropriate category under paints. After selecting the paint category, the cursor is hovered around the required paint under the sub-category followed by clicking on the arrow to add the required paint to the set of materials in the current model. The procedure can be repeated to put additional colours of paint as desired. Each paint that has been added can be renamed by right-clicking on it and selecting **Rename** then giving it an appropriate name. Renaming the materials/paints for the current session does not change the names of paints stored in the database but simplifies the process of adding the paints to the various walls.

Assuming that the user will be modelling a building with different walls such as the extrusion in Figure 9.1 and that the required number of paints/materials have been added to the current session and renamed, the next step would be to add these materials to the wall surfaces. The AutoCAD command SOLIDEDIT allows users to edit edges and faces of 3D solid objects, almost equivalent to exploding the object as discussed in earlier chapters. However, the command cannot be used with mesh objects but may be required to convert it to a 3D solid first. Using this command, followed by Face and then Material on the set of options will enable the user to select faces. Once a face has been selected, then enter to prompt for the name of the material for that particular face, which can be typed from the renamed paints in the current session and entered again to accept the materials edit. This will add the material only to the selected face or wall. Once the materials have been added, the process can be repeated for the other walls. These are basically two key ways of adding materials to any 3D solid object or faces, but there are other ways of doing the same such as defining layers for a session and attaching materials to the different layers etc.

9.7 PRACTICAL APPLICATIONS FOR 3D MODELLING

Being the ultimate representation of objects in 3D, solid models are used for a wide variety of applications despite them being more involving to create compared to wireframe and surface models. CSG is used in cases where simple geometric objects are required or where mathematical accuracy is important. Almost all CAD systems employ CSG technology especially for representing tool cuts in manufacturing, and features where parts must fit together in machine assemblies.

9.7.1 CSG models

CSG technology has been used for the Quake and Unreal engines as well as the Hammer, and Torque Game Engine/Torque Game Engine Advanced because of the ability to use relatively simple objects such as boxes, cylinders and spheres to create geometrically complex objects (Zuo et al., 2014) as demonstrated in this chapter. When it is used in procedural or parametric modelling, CSG enables them to revise any complex geometry by changing the position of objects or the Boolean operations used to combine them. CSG is also advantageous in that it can easily assure that objects are 'solid' or water-tight if all of the building blocks or primitive shapes are water-tight (van Rossen & Baranowski, 2011). This is useful, particularly for some manufacturing or engineering computation applications.

In comparison to creating models using B-Rep, additional topological data are required, or consistency checks must be performed to assure that the given boundary description specifies a valid solid object. Another important advantage is that CSG shapes are easier to use to classify arbitrary points as being either inside or outside the shape created by CSG. The points can conveniently be classified against all the underlying primitives and the resulting Boolean operations worked out. This is a desirable and useful quality for some applications such as ray tracing (Vasiou et al., 2018).

CSG models are often better than meshes because of their parametric nature and the simple shapes used to construct them, especially when developing customized models. For such customized applications, it is useful to convert already existing meshes to CSG trees. Operations such as SOLIDEDIT would not be possible on meshes but they have to be converted to 3D solids first, sometimes referred to as inverse CSG.

CSG trees created from this are required to occupy the same volume in 3D space as the input mesh while having a minimal number of nodes. In cases like this, simple solutions are necessary to ensure that the resulting model can easily be adjusted. Sometimes this is also problematic because of the large search space that has to be explored. It combines continuous parameters such as the dimension and size of the primitive shapes, and discrete parameters such as the Boolean operators used to build the final CSG models.

This challenge is usually resolved by deductive methods through building sets of half-spaces that describe the interior of the model geometry. Such half-spaces are used to describe primitives that can be combined to get the final model (Buchele & Crawford, 2004). An alternative way is to decouple the detection of primitive shapes and the construction and computation of the CSG tree that defines the final model. This approach makes use of the ability of modern program synthesis tools to find a CSG tree with minimal complexity (Du et al., 2018).

Genetic algorithms (GA) have also been used as an alternative approach to iteratively optimize an initial shape towards the shape of the desired mesh (Fayolle & Pasko, 2016). The modelling of 3D objects is commonly derived from a shell/boundary approach. A hybrid framework between the interactive genetic algorithm (IGA) and the L-system has also been used to steer 3D modelling using subjective preference feedback from users in which the creative processes in the proposed hybrid system were presented for generating models using the approach (Kiptiah binti Ariffin et al., 2017).

9.7.2 Rendered images

Rendered images have uses in architecture, video games, simulators, movie and TV visual effects and design visualization. Each of these makes use of a different balance of features and techniques. Despite a lot of time and space required to render images to realistic levels, a wide variety of renderers are available for use. Some of these are integrated into larger modelling and simulation packages, some are stand-alone and some are free open-source projects. Rendering is a carefully engineered technique based on multiple disciplines, including mathematics, physics, visual perception and software development.

9.7.3 3D printing

3D printing, commonly referred to as additive manufacturing is a recent technology developed for making 3D solid objects from electronic files in conjunction with computer-aided manufacture (CAM). It is a combination of CAD and CAM that works in line with the principles of solid modelling covered in this chapter. The creation of 3D printed objects is achieved using additive processes to develop realistic or life-like 3D objects. In additive manufacturing, an object is created by laying down successive layers of material until the desired object is built. Each of these layers represent thinly sliced cross-sections of the object (Candi & Beltagui, 2018).

3D printing technology can be viewed as the opposite of subtractive manufacturing which involves cutting out pieces of metal or plastic as carried out CAM but with the same ultimate result of building the desired object. Thus, 3D printing enables the production of complex shapes using less material than traditional manufacturing techniques that in fact remove material to get to the desired object. The basis of operation for 3D printing is the 3D solid models that were developed in this chapter using B-Rep, CSG or solid of revolution.

There is a wide variety of software tools for 3D printing, ranging from industrial (commercial) to open source. Some of these include TinkerCAD which was discussed briefly in Chapter 1. This is readily available software that can be used in conjunction with most web browsers. The package has several options to export developed electronic models for printing and these include printable files such as Standard Triangle Language (STL) or Object (OBJ) files. Once the STL or OBJ printable is ready, it is then 'sliced' to prepare it for a 3D printer. This involves slicing up the 3D model into hundreds or thousands of layers, also done by specialized slicing software and then finally fed to the 3D printer through USB or Wi-Fi for printing layer by layer, hence 'additive manufacturing'.

3D printing has been popularized over recent years in this era of the Fourth Industrial Revolution as more and more manufacturing industries are adopting the technology. Whereas 3D printing was only suitable and used for physical prototyping and one-off manufacturing in the early stages, it is now rapidly transforming into a regular production technology. The demand for such technology has become industrial in nature and it is projected that the global 3D printing market will reach $41 billion by 2026 (Candi & Beltagui, 2018).

It is also projected that 3D printing technology will transform almost every major industry and change the way people live and work as the world gradually moves towards the Fifth Industrial Revolution (Digital Ecosystem). 3D printing basically involves many forms of technologies and materials currently being used in industries throughout the world. It can thus be viewed as a cluster of diverse industries with a myriad of different applications (Sivasankaran & Radjaram, 2020). Examples of 3D printed objects include consumer products such as eyewear, footwear, design, furniture, dental products such as false teeth, industrial products such as manufacturing tools, prototypes, functional end-use parts, prosthetic limbs, architectural scale models of buildings, replications of ancient artefacts and many other models that can be created by additive manufacturing including 3D printed houses.

Progressive industries have now invested in 3D printers for their designs to create prototypes since the late 1970s because of their speed and low cost compared to manual prototyping. Iterations and any other changes and permutations are so much easier and cheaper to make, without the need for expensive moulds and tools traditionally used in toolmaking. Other than rapid prototyping, 3D printing is also used for rapid manufacturing in the batch production of many consumer goods. 3D printing has thus become a new form of rapid manufacturing where businesses use 3D printers for short-run/small-batch custom manufacturing.

9.8 SUMMARY

Several techniques can be employed to visualize and analyse objects in three dimensions. This chapter covered the main ones that include B-Rep, solids of revolution and CSG. The major difference and advantage of solid models over wireframe and surface models covered in Chapter 8 are that they describe objects completely in terms of their shapes, surfaces and materials. Although solid models are difficult to generate, the amount of information they contain allows better visualization and analysis. This chapter also covered the various visual styles that can be used to provide a better understanding of objects and models. In addition, the attachment and specification of materials to objects or faces of objects allows for the determination of such properties as mass moments of inertia, weight as well as checking for interferences in assemblies where parts fit into each other. Apart from the guidelines to effectively create solid models and use them, the chapter also covered practical applications for 3D solid models. These include the use of CSG models in procedural and parametric designs and models, rendered images that represent realistic replicas of objects on computer as well as 3D printed objects based on additive manufacturing. The next chapter will focus on these tools in a practical and worked-out example to demonstrate the usefulness of 3D technology.

9.9 REVIEW EXERCISES

1. Illustrate using two schematics, the fundamental parameters required to produce surface and solid models of revolution using B-Rep and CSG.
2. Outline the philosophy behind the mathematical logic for Boolean algebra and how this concept is applied in CSG, using at least one model that makes use of all six solid primitives as well as the three Boolean operations.
3. Three-dimensional models can be viewed in CAD systems using different visual styles. Explain the differences by defining the following: Realistic, Conceptual, Shaded, Sketchy and Rendered.
4. List and explain, stage by stage, the process of assigning materials to 3D solids, focusing on entire objects and parts of objects such as faces.
5. Outline at least three of the major practical applications for 3D solid models.

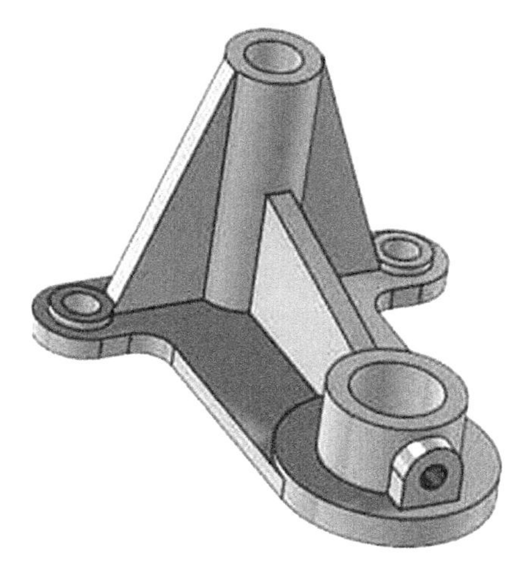

Figure 9.9 Pictorial view of a cast iron bracket.

6. Explain the relationship between 3D solid models and 3D printing, focusing on subtractive and additive manufacturing.
7. Figure 9.9 shows a pictorial view of a cast iron bracket. List the optimum number of solid primitives required to produce it, then reproduce it in AutoCAD using CSG. Estimate measurements for your model.

Chapter 10

3D solid modelling in assembly and analysis

10.1 INTRODUCTION

Having assimilated and grasped the techniques involved in solid modelling from boundary representation (B-Rep) and extrusions, solids of revolution to constructive solid geometry (CSG) in general, it is now important to put these to practical applications. This chapter will focus on two case studies, the first of which will be presented in the form of a tutorial, a step-by-step worked-out example for a wheel base assembly that consists of several components of different materials such as steel cast, bronze and mild steel. The tutorial will guide users, not only on how to model the various solid elements and put them together in an assembly but will also go further to define and specify materials for the components as well as carry out some analyses such as checking for interferences and mass moments of inertia.

The second will be a case study carried out on a failed connecting rod for a diesel engine for a transporting company. The connecting rod also consists of several components, albeit mostly of the same cast iron but this will be used to demonstrate the use and usefulness of CSG in engineering design and analysis. The two case studies enable AutoCAD users to effectively use their CAD systems to the maximum, in tandem with global changes and competitions as the world gradually transitions towards the Fifth Industrial Revolution.

10.2 MODELLING WHEEL BASE ASSEMBLY: TUTORIAL

The wheelbase assembly used in this tutorial comprises five components as shown in the cross-section in Figure 10.1. The sequence adopted in this tutorial is first to model generate solid models of the individual components, during which the materials are also specified. This is followed by assembling the components and then visualization, analysis and lastly automatic conversion of the 3D solid model of assembly to any required 2D orthographic elevations. This tutorial combines the knowledge and skills acquired throughout this book from simple construction of building blocks such as

DOI: 10.1201/9781003288626-10

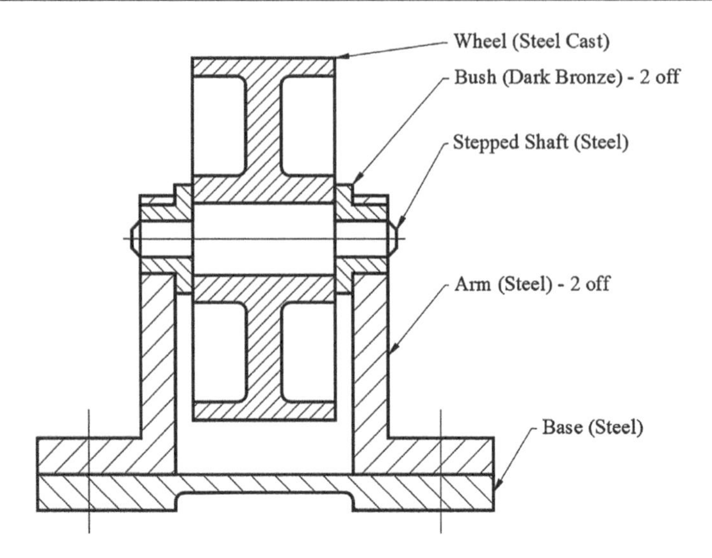

Figure 10.1 Sectioned assembly of the wheelbase.

lines and polylines, use of layers, 3D modelling, analysis and automatic generation of 2D orthographic views from 3D models. The components can be generated in the default plan view (0, 0, 1) but once they are revolved, they will be difficult to visualize hence it is best to set the viewing point to a suitable 3D view such as (1, –1, 1) which is the SE Isometric or any other suitable viewing angle. For ease of analysis and management of the model, all components will be modelled on their own layers.

10.2.1 Modelling the wheel

The first step would be to define the layer *Centre*, with the default colour (black or white depending on what the background of the screen is) and linetype **Centre**, then draw the construction line using any of the options of *Absolute*, *Polar* or *Relative*, but noting that usually the first point has to be *Absolute* and the rest can be constructed using *Polar* of *Relative*. In this tutorial, all such coordinate systems will be made use of, for familiarity. Starting the Centre line at Absolute, **(0, –2)**, for example, then terminate it 4 units along the *y*-axis, **@0, 4, 0** (*Relative*). If the line is not visible, Zoom Extents can be used to view the entire line on the screen. Similarly, if the centre line is not visible still, the LTSCALE variable will need to be adjusted according for better visibility.

The next step would be to define the second layer, *Wheel* with a continuous linetype and red colour. The central opening (hole) for the wheel is 1 unit in diameter, hence the path curve will be offset from the construction centre line by **0.5**. The path curve for the wheel, which will eventually be revolved should be constructed using polylines, using the dimensions

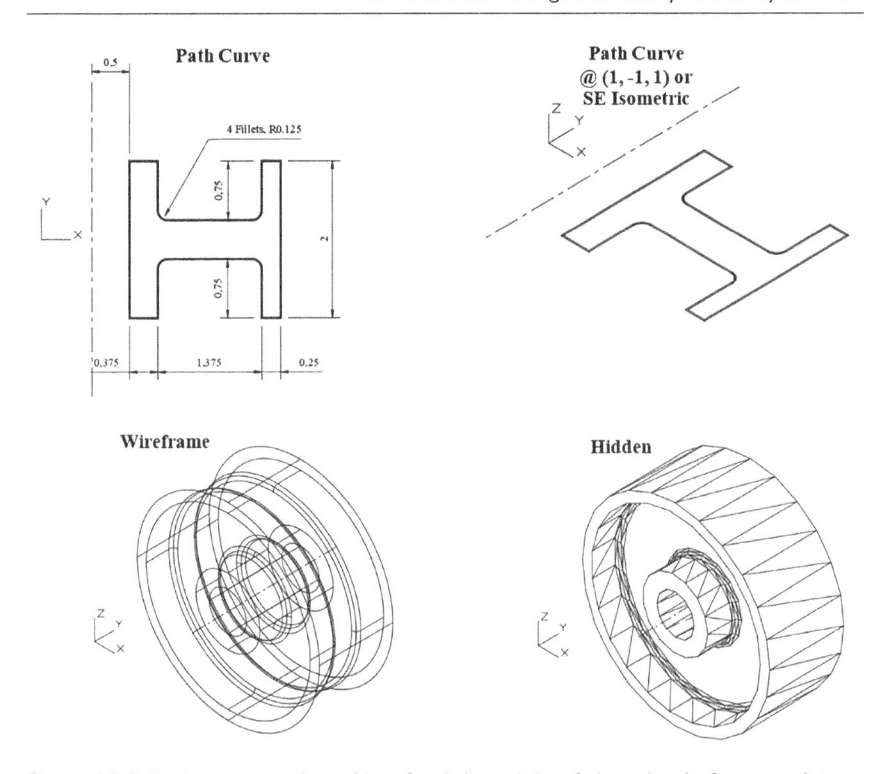

Figure 10.2 Path curve and profile of solid models of the wheel after revolving.

provided, starting at *Absolute* point (0.5, –1) all the way round back to the same starting point. If the PLINE command is used to generate the polyline from beginning to end, then it will be one continuous but if for some reason the user had to break and restart the command, then all the polylines must be joined together using PEDIT to form a continuous polyline for the required path curve. The internal webs of the wheel are at fillet radii of **0.125** and can be inserted using the FILLET command as covered in Chapter 3 until all the four fillets are done and as shown in Figure 10.2. At this point, the pat curve for the wheel is complete and the material, steel cast can be selected from the material browser to select the path curve to define its material. Finally, the REVOLVE command is used to obtain the 3D solid model in Figure 10.2 (both hidden and wireframe, together with the dimensioned profile and 3D view of the profile). This can be saved as **Solid Wheel.dwg**.

10.2.2 Modelling the shaft and fitting to wheel

The central shaft will also be created by the solid of revolution similar to the wheel. To avoid a cluttered screen, the solid model of the wheel can temporarily be removed by turning off the *Wheel* layer. A new layer for the shaft

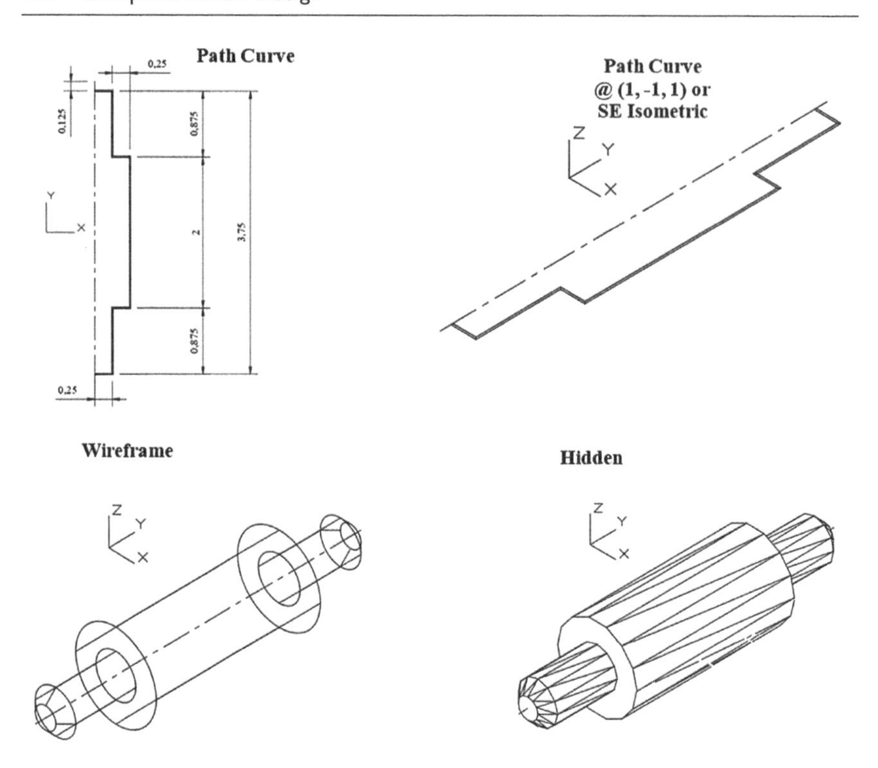

Figure 10.3 Path curve and profile of solid models of the shaft after revolving.

is then added with a different colour, say magenta, continuous linetype and named *Shaft*. Using the construction centre line, a continuous polyline that will constitute the path curve of the shaft can be constructed, as shown in Figure 10.3, offset from the top end of the centre line by a small amount of about 0.125, from where the polyline can start until the bottom end of the construction line, using the dimensions provided in Figure 10.3. The chamfers, whose first and second distances are both 0.125 can be constructed on the path curve or after revolving the shaft to form a solid shaft, which can be saved as **Solid Shaft.dwg**. Using the material browser, the shaft is defined as steel.

10.2.3 Modelling and assembling the bushes

As can be seen from Figure 10.1, there are two dark bronze bushes in the wheel base assembly. Before modelling and constructing the bushes, the two layers, *Wheel* and *Shaft* are turned off leaving the *Centre* layer ON, to avoid cluttering the screen and also to aid in the construction of the bush. A new layer Bush is then created and given a different colour from the previous ones, say, green. The path curve of the bush can be created anywhere adjacent to the centre line, as long as it is offset from the centre line by a distance

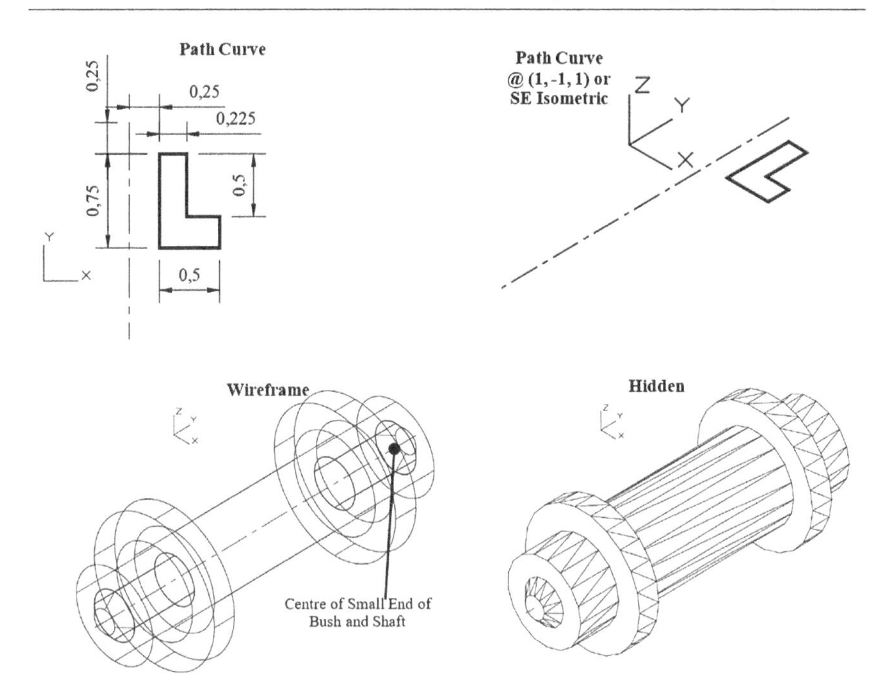

Figure 10.4 Path curve of bush and solid models of the bush and shaft.

equivalent to the radius of the mid-section of the shaft (0.5), over which it will be fitted tightly in assembly. Using the dimensions given in Figure 10.4, the path curve should also be created using polylines, that are continuous or can be joined to form one continuous polyline in case the user is interrupted along the way. After the path curve has been constructed, the solid bush can also be obtained by revolving the path curve using the centre line as the axis of symmetry.

So far, the bush is the only component that has been created anywhere along the axis of symmetry but not in the position it is supposed to be in the assembly. The wheel and shaft have both so far been placed in their correct positions of assembly. At this stage, the assembling of the wheelbase can begin by moving and positioning the bush correctly. However, in order to do this precisely, the *Shaft* layer must be turned on in order to use it as a reference for the positioning of the bush. The MOVE command can then be used, followed by selecting the bush and for the base point, the object snap point Center is used to pick the centre of the smaller end of the shaft and for the second point, the object snap point Center is also used to pick the centre of the small end of the stepped shaft, where the chamfer starts. This is the point where the bush is supposed to sit according to the sectioned assembly in Figure 10.1. For the movement of solids from one position to another, previous versions of AutoCAD used the SOLMOVE command followed by specifying whether to translate or rotate it along or about any given axis, for example, ty100 meant to move the object from a chosen point along the *y*-axis by 100 units etc. The newer versions of the software simply make use of the normal move (M) command followed by appropriate object snap points.

At this point, if the HIDE option is chosen, the assembly may not come out clearly because of the other layers that are turned off. Depending on the position, it might also be useful to freeze layers that will not be in use at this point. At this point, the material for the bush, DARK BRONZE can be selected using the Materials Browser. If this particular material does not come up under the usual categories of metals, it can be searched for in the AutoCAD materials database by typing in BRONZE at the top of the Materials Browser and all available bronze materials are then listed from where the dark bronze can be selected. As with the wheel and shaft, the DARK BRONZE can be dragged from the Materials Browser and set to define the bush. The materials browser will also show at this stage that there are three materials in the assembly, Cast Steel, Steel and Dark Bronze.

From Figure 10.1, it is also evident that there are two bushes on the wheel base assembly, one on each base of the shaft. These do not need to be created individually as the other one is a replica of the other. As such, the mirror command (MI) can also be used to mirror 3D solids such as the bush. For the objects to be mirrored, the bush that has just been assembled can be selected. This needs to be mirrored along the axis that is at right angles to the construction centre line. As such, for the first point of the mirror line, the object snap mode mid-point can be used to pick the mid-point of the construction centre line and for the second point, an arbitrary point can be selected on either side of the centre line but with orthogonal ON (selected while holding down the SHIFT key). That way, the other bush will thus be placed in a mirror position on the other side of the shaft. The model can then be saved as **Solid Shaft and Bushes.dwg** at this point.

10.2.4 Modelling and assembling the arms

The next components to be modelled and assembled after the wheel, shaft and bushes are the two arms on both ends of the wheel. The arms consist of the upright wing and the arm base. The layers for the three components that have been modelled and assembled so far, together with their dimensions can now be turned off to avoid cluttering the screen, thereby speeding up the modelling. A new layer, *Arm* can be created with a different colour, say red and it is made in the current layer to enable the construction of the arm components under that layer. As shown in Figure 10.5, the upright wing of the arm consists of polylines, an arc and a construction circle where the cylinder to form the hole for the bush will be inserted.

The profile or path curve of this upright wing of the arm can be constructed using the dimensions provided in the first drawing in Figure 10.5. The position of this wing model does not matter much at this stage as it will

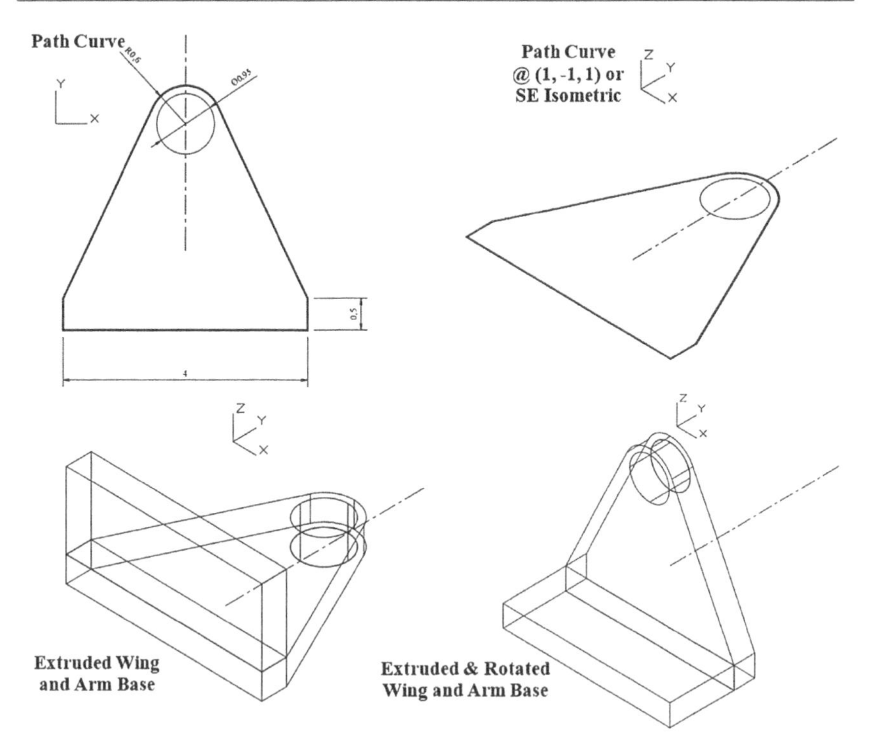

Figure 10.5 Path curve of arm wing and extruded and rotated model.

be rotated and moved into its correct position during assembly. First, the circle of diameter 0.95 is drawn followed by another concentric circle of diameter 1.2 then the bottom lines of the arm wing, ensuring that they are positioned correctly in line with the circles and dimensions provided. The last two slanting lines will then be drawn using the object snap mode, with endpoints of the two bottom vertical lines to the tangent of the larger circle. The larger circle is then trimmed to remain with an arc as shown in Figure 10.5. Using PEDIT, the lines and arc are then joined to form a continuous polyline, which will constitute the path curve for the solid wing of the arm. At this stage, the materials browser can be opened and steel is assigned to this wing arm. As has been the practice, the models should be periodically saved using appropriate names, in this case, it can be saved as **Arm Wing Profile.dwg**. An appropriate 3D view such as SE Isometric (1, −1, 1) can be used to view the path curve in 3D, as shown in the second drawing on the top right of Figure 10.5. Depending on the width of the polyline chosen or defined, the 3D view path curve may appear as two parallel lines but this does not matter for the purposes of extruding the profile.

Using the EXTRUDE command, the arm wing is then extruded upwards to a height of 0.5. It must also be noted that when extruding objects, before stating the height, it is important to guide the system by showing in which direction the extrusion should go, using the mouse. In order to fit the arm base as shown in the bottom left model in Figure 10.5, the BOX command is used and the extruded upper right corner and the other diagonal opposite corner are used as first and second corners and a box height of 1.5 to give the solid models shown in the Figure 10.5 (bottom left). The construction circle can be used to generate a cylinder of diameter 0.95, through which the bush will be assembled. In order to create the counter-sunk or spot-faced holes as shown in Figure 10.6, the model needs to be rotated in 3D in order to stand upright, as if ready for assembly. The command ROTATE3D is used to select the arm and its base followed by selecting the x-axis through which it will be rotated by 90° anticlockwise.

At this stage, it is not critical to select a particular point on the x-axis but any object snap on the model will do. The 3D rotated model appears as shown in the bottom right of Figure 10.5.

The counter-sunk holes on the arm base will be created by cylinders which will later be subtracted from the united model. The smaller hole is generated by first creating a cylinder of diameter, 0.39 at the bottom, to a height of 0.5, equivalent to the thickness of the arm base. The location of the cylinder can be derived from the dimensions provided in Figure 10.6. The second and larger hole is of diameter 0.875 and is created in the same manner. However,

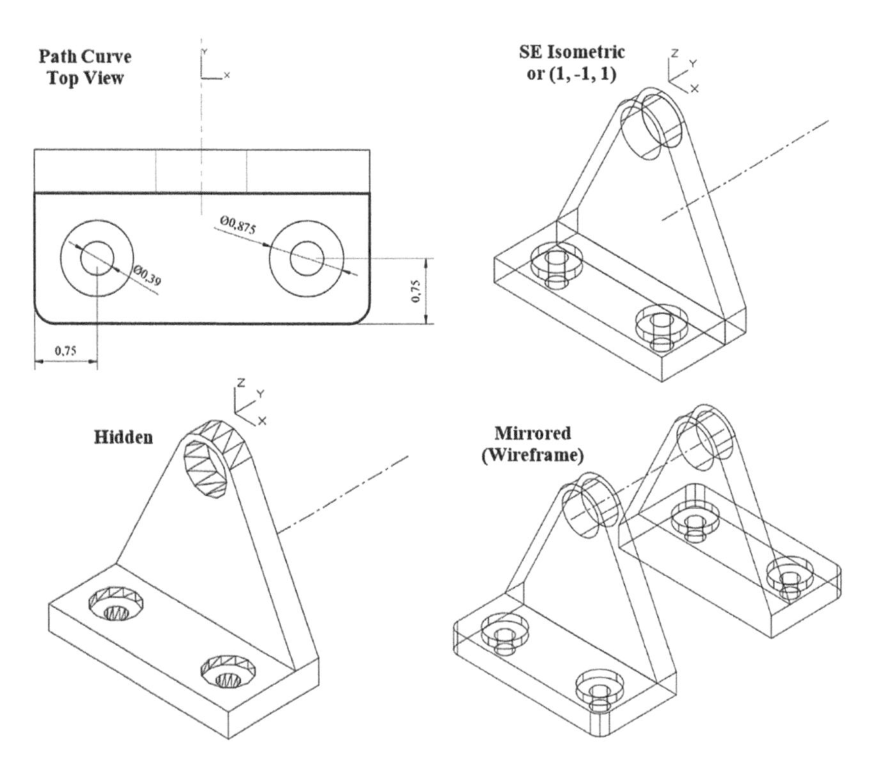

Figure 10.6 Complete arm with counter-sunk holes: path curve, rotated and mirrored.

the centre of this cylinder will be at the top, to coincide with the centre of the top end of the small cylinder. Hence, before specifying the height, the mouse can be used to guide the system by a height of 0.175. The next step would be to make all the parts of the arm and its base, one unit. The other three countersunk holes can be created in the same manner but it is easier to use a mirror to create the one adjacent to the first one and then mirror the two to the other side. This is accomplished by first uniting the arm wing and the arm base, then subtracting the five cylinders (the one on the arm wing and the four counter-sunk on the arm base) as shown in Figure 10.6.

The last step in the modelling of the arm would be to assemble it by moving it into the correct position of the assembly. This is done by first turning on the layer for the bush and then moving the arm using the object snap point Center of the larger cylinder on the arm wing to the corresponding Center of the bush in the assembly. That way, the united arm and its base will now have been assembled. Fillets of radius 2.5 can then be added to the two front corners of the wing base as shown in Figure 10.6. At this stage, only one arm on the left was created. The other arm can be created using the mirror (MI) command where the original arm is selected, followed by specifying the mid-point of the construction centre line. The second point can be defined by using the mouse and watching that the second arm will be in the right place, but more importantly that orthogonal must be on by holding down the SHIFT key as the second point is selected. The model can be saved as **Solid Arm.dwg** at this stage.

10.2.5 Modelling and assembling the base

The last component of the wheel base assembly is the base itself, upon which the entire assembly developed so far, will be mounted. A new layer, *Base* is created with a different colour, say, yellow, which is then made into the current layer while all the other layers are turned off to avoid cluttering the screen. Using the dimensions provided in the top right of Figure 10.7, the path curve or profile of the base is generated. This consists of a series of lines and two arcs which must all be joined together using PEDIT to form a continuous line. The path curve can be created in any position or orientation as the resulting solid model will eventually be moved and rotated twice before it is placed in its correct assembly position. To avoid losing any details, the base model can be saved as **Base Profile.dwg** at this stage.

Using the materials browser, the base material can be set to steel. The model of the base thus far can be set to a suitable 3D view such as SE Isometric (1, −1, 1) as shown in the top right corner of Figure 10.7. The entire profile can then be extruded to a height of 6.5, derived from the dimensions provided in Figure 10.7. Before adding the fixture holes on the base, the extruded base is first rotated in 3D about the x-axis through 90° anti-clockwise using the ROTATE3D command to position it as shown in the bottom right corner of Figure 10.7.

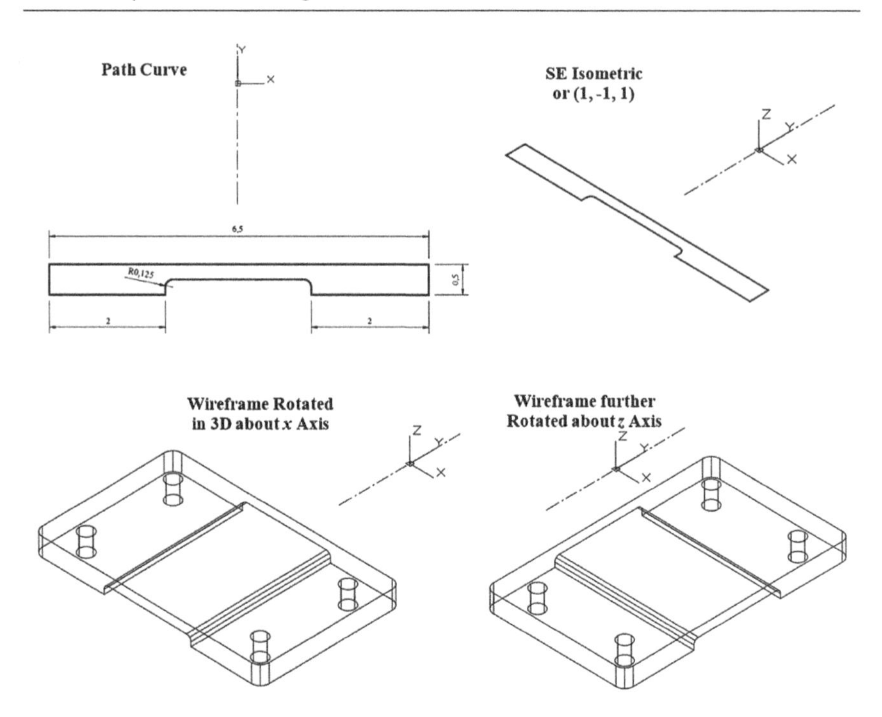

Figure 10.7 Assembly base path curve and wireframe model.

At this stage, and in conjunction with the dimensions provided in Figure 10.6, the four cylinders of diameter, 0.39 each can be added to the rotated model, using the same procedure as in the arm base. Only one of them needs to be created and the other is obtained by mirroring the first one to the appropriate position, with the orthogonal state on (holding down the SHIFT key, in the mirror process).

The four cylinders are then subtracted from the base to form the fixture holes. In addition, the four corner fillets of radius 0.25 each can be added to each of the four corners of the base. Although the model is complete, it should further be rotated in 3D about the z-axis through 90° anticlockwise in order to orient it correctly before moving it into the correct position of assembly. In order to move the solid base into its correct position of assembly, the *Arm* layer can be turned ON in order to use object snap points to link the base to the assembled arms. Using the move (M) command, the centre of the top end of one of the cylinders on the base can be used as the base point while the centre of the bottom end of the matching cylinder on one of the arms will be the second point of displacement. This will accurately assemble the base into its correct position of assembly. Due to the small sizes of the cylinders, it might be useful to zoom in on the area containing the two corresponding cylinders in order to adequately pick the centres without the clutter of lines and circles. Figure 10.8 shows the assembled model so far, in its wireframe and hidden visual styles, while Figure 10.9

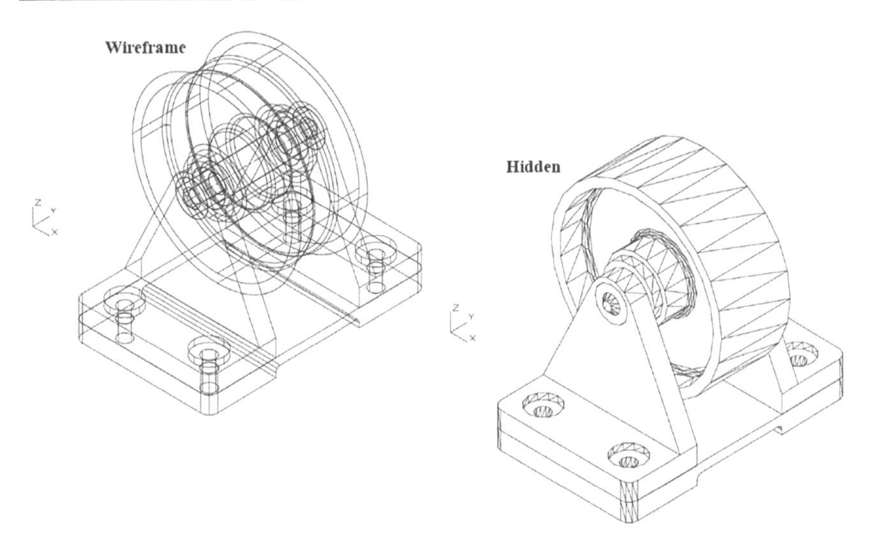

Figure 10.8 Complete wireframe and hidden models of the wheel base assembly.

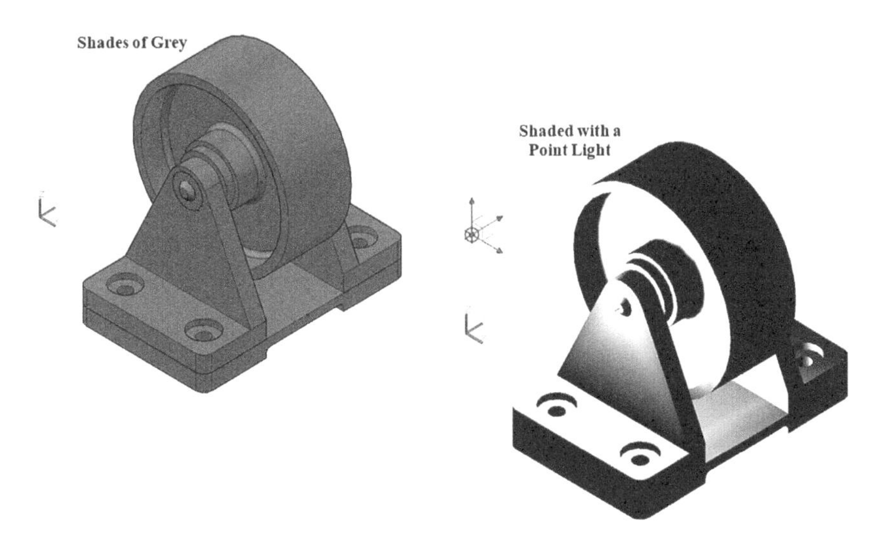

Figure 10.9 Complete model of the wheel base assembly shaded with point light.

shows the more realistic visual styles of grey and shaded with a point light illuminating one of the faces.

10.3 ANALYSIS OF THE SOLID MODEL

Having completed putting the components of the wheel base assembly together, there are various tools that can be used to analyse the assembly during which time, if any anomalies are identified, they can be corrected to ensure that the right model has been developed. Such analysis ranges from cutting open the assembly to see how the internal parts fit, checking for interferences of parts that fit together, determining mass properties and automatically creating orthographic views from the solid model of the assembly.

10.3.1 Sectioning model of the assembly

The model can be sectioned along any plane to open it up in order to view the setup of components within the assembly. This is particularly useful when there are many components. Previous and earlier versions of AutoCAD (up to Release 12) used the SOLCUT command but the later versions (from Release 12 on) now use the SLICE command.

There are four parameters that need to be defined before a model can be sliced or sectioned. The first is to select the solid objects to be sliced, and in the case of the wheel base assembly, all five components including the duplicates for bush and arm that were created by the mirror, need to be selected. Users need to be careful because of the clutter of wireframe lines and, circles and arcs, that all components are selected. It may be necessary to zoom in on particular areas to be able to pick some of the components that may be 'buried' behind others.

The second parameter is to specify the cutting plane, which also has several options but for this tutorial, the default three-point option will be used to specify any three points on the cutting plane, preferably making use of the object snap point, midpoint to pick the midpoints along any of the components as shown in Figure 10.10. Once the three points are selected, the system

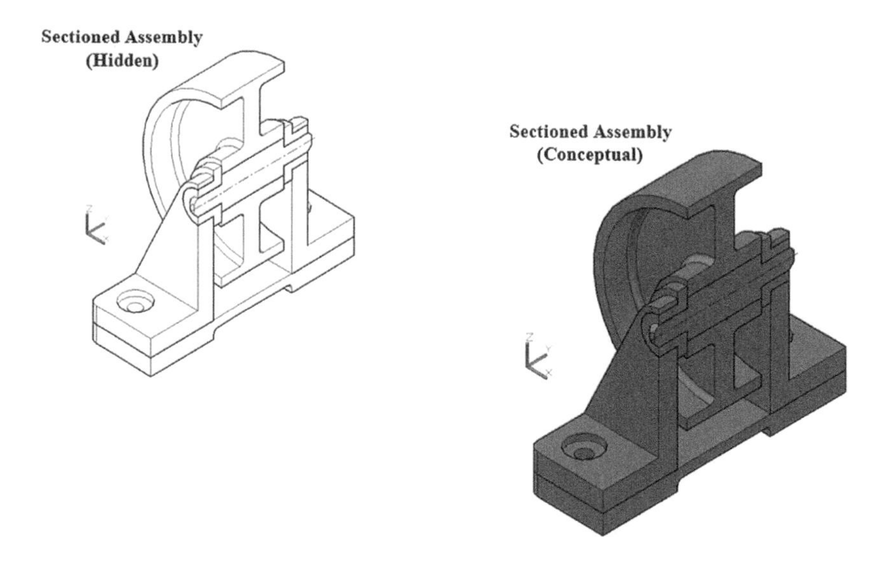

Figure 10.10 Sectioned assembly of the model in hidden and conceptual visual style.

will automatically prompt the user to pick any point on the desired side of the model. The desired side of the model will be the section that the user wishes to retain and in this case, any point at the back of the model will suffice. However, before selecting that point, if the user wishes to retain both sectioned parts, then the option BOTH should be selected, otherwise, just picking a point on the desired side will automatically slice and remove the undesired part, leaving the display in Figure 10.10 in any desirable visual style.

10.3.2 Checking for interferences

The AutoCAD 3D modelling facility also allows users to check whether any parts that are fitting into each other are interfering in any way. Modelling in AutoCAD is electronic and not like physical prototyping where the designer can physically observe a problem with fitting components when developing the physical prototype. However, with CAD systems, components can be placed anywhere, hence the need to ensure that none of the parts is sitting in interference with others within the assembly. In most cases, such interferences are detected by the CAD systems and reported where components share volumes.

The AutoCAD command to check for interferences is INTERFERE, followed by selecting the objects that need to be checked. In the case of the wheel base assembly, almost all the components fit into each, starting with the shaft in the middles, over which the bushes are fitted, then the arms fit onto the bushes and wheel over the shaft. Therefore, such components can be selected or simply selecting a window of the entire assembly.

Once the selections are complete, the command is terminated by entering (↵) and the default option that follows would be to check by entering (↵) again. If all the parts are sitting properly without sharing a volume with the others, the system will report back 'Objects do not interfere'. However, if there are interferences, this will also be reported, prompting the user to recheck and correct any mistakes that may have been encountered during assembly. Sometimes when parts that fit into each other and have exactly the same measurements, such as the diameters of the bush and shaft, that of the arm and bush, as well as the wheel hole and the central part of the shaft, the system will prompt that number of solids interfering. Normally such parts that fit into each other should be left with a very small tolerance for a clearance fit. Where such errors of interference occur, the user is given the opportunity to recreate or modify the solid components to ensure that there is no interference. Similarly, for intersecting volumes, the solid models should be edited to correct the errors.

10.3.3 Determining mass properties

Engineering designers require pertinent information for decision-making during the design and development stages. For physically prototyping, such information can physically be measured or calculated using different measuring instruments. However, for electronic prototyping such as in 3D CAD solid modelling, that information can be calculated and availed to the user through MASSPROP command. This command simply requires the user to select a particular solid or an entire assembly and automatically, various properties of the selected solid will be displayed, depending on what units and materials would have been set during the development stage. The material analysis can be saved to an electronic file, **Name.mpr** containing the mass properties, that can be accessed using Notepad. An example of this is shown in Figure 10.11, extracted from the tutorial on the wheel base assembly.

The system calculates and avails such information as the mass and volume of selected solids and the bounding box size in terms of (x, y, z) coordinates. The other typical information necessary for design engineers, included in the mass properties files, includes the centroid (centre of gravity for the selection, moments of inertia, products of inertia, radii of gyration as well as principal moments and X-Y-Z directions about the centroid.

10.3.4 Automatically generating orthographic views

For the purpose of detailed orthographic drawings, which are normally used in production or on construction sites, as described in Chapter 6, these can

```
Complete Wheel and Base.mpr - Notepad
File  Edit  Format  View  Help

 ----------------   SOLIDS    ----------------

Mass:                    45.6345
Volume:                  45.6345
Bounding box:         X: -0.8200  --  4.1800
                      Y: 1.5578  --  8.0578
                      Z: -3.7500  --  2.5000
Centroid:             X: 1.6800
                      Y: 4.8078
                      Z: -1.5223
Moments of inertia:   X: 1406.4250
                      Y: 435.8893
                      Z: 1349.9764
Products of inertia: XY: -368.5975
                     YZ: 333.9818
                     ZX: 116.7065
Radii of gyration:    X: 5.5515
                      Y: 3.0906
                      Z: 5.4390
Principal moments and X-Y-Z directions about centroid:
                      I: 245.8541 along [1.0000 0.0000 0.0000]
                      J: 201.3403 along [0.0000 1.0000 0.0000]
                      K: 166.3493 along [0.0000 0.0000 1.0000]
```

Figure 10.11 Typical mass properties file opened using notepad.

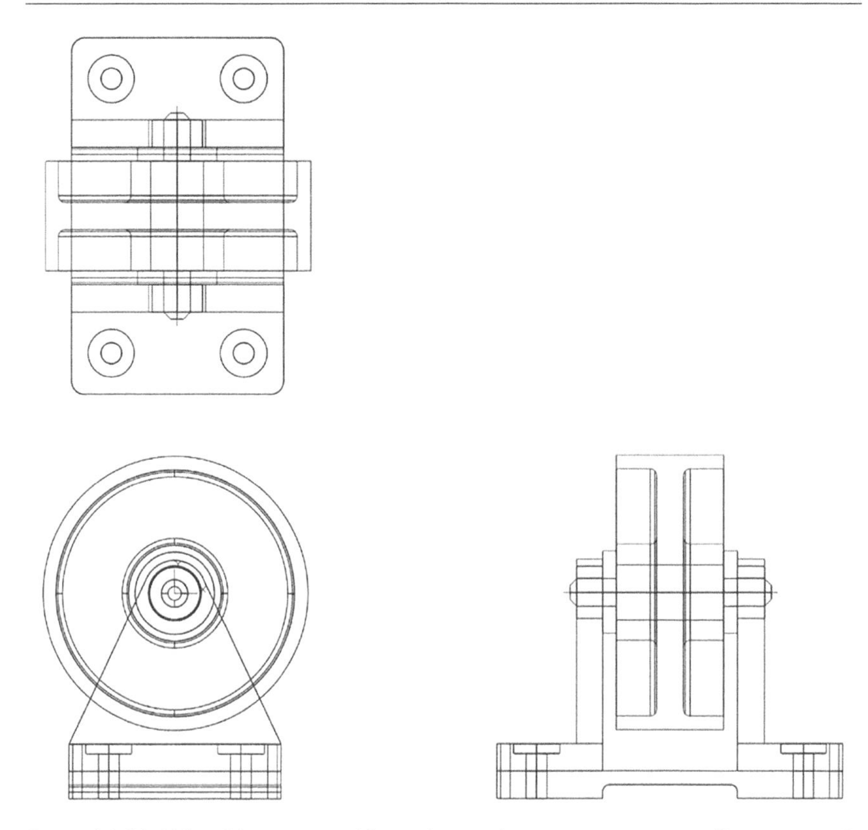

Figure 10.12 Wheel base assembly orthographic views automatically generated.

be generated automatically once all the 3D modelling has been completed. With the solid model on display, the **View** pull-down menu can be used, followed by **3D View** to choose an appropriate orthographic projection such as Top, Bottom, Left, Right, Front or Back. The selected views can be saved and organized into a properly detailed engineering drawing, as shall be discussed in detail in Chapter 12 under management of drawings using viewports. After obtaining required orthographic views such as those in Figure 10.12, the user can refine them by first exploding and then detailing them by inserting appropriate hidden lines as well as dimensions.

10.4 MODELLING CONNECTING ROD FOR A DIESEL ENGINE: CASE STUDY

The data, modelling and analysis in this section are a summarized version of a research/case study carried out on a connecting rod of a diesel engine for a heavy vehicle maintenance and service company (Nyemba, 2000), which was subsequently published as a paper in the *African Journal of Science and Technology*, from where more details can be obtained. The use of 2D

Figure 10.13 Photograph of a bend and failed connecting rod.

drafting as a modelling tool is assumed to be well understood as it is the base on which engineering and technical drawing are taught, but the potential of 3-dimensional modelling is much less exploited. This case study was derived from a heavy vehicle maintenance and service company whose diesel engines frequently failed. Upon opening the engines, some of the connecting rods would be bent as shown in Figure 10.13. The connecting rods were modelled and eventually analysed using ANSYS Finite Element to determine the source of faults in operation. However, this section focuses on and is limited to the modelling aspects of the connecting rod using CSG, as a contribution to the book.

Models obtained by using such methods can be quite beneficial to the designer at the conceptual stage of design or redesign because they can be used to modify operating characteristics or the behaviour of the real objects. The same models can be exported to Finite Element Modelling systems for stress and strain analysis under loading, as well as properties such as the weight, moments of inertia, etc. can be extracted and used in the design process.

Modelling is a word like design, which is used rather generally, in everyday language. This can range from the display of fashion and style to the construction of scale models that represent real objects, in all cases trying to represent something realistic. Modelling can be a much more formal activity when a design engineer or architect wishes to explore the nature of some physical process or understand how to construct or operate some physical object or system (Remondino & El-Hakim, 2006).

The analysis of engineering objects prior to physical prototyping is not only cost-effective but allows designers to understand more of what they are dealing with in terms of their behaviour and performance under function. Therefore, care has to be taken to adequately represent the real thing (modelling) as models are only substitutes for the actual physical objects. As alluded to in previous chapters, objects can be represented in many different ways, and it is up to the designer to select the most appropriate has its modelling tool has its own characteristics, which are useful to the designer, such as; manufacturability, aesthetics, and the shape in general.

Models are usually constructed from discrete parts, even though a designer may think of the real object as being continuous and infinitely divisible.

To model the connecting rod, there was a need to focus on those aspects which can be represented formally and which can be considered relevant to the task in hand. These aspects tended to be geometrical, physical and material properties. The most important characteristics relevant to the functioning of a connecting rod are; shape, motion, force and torque, temperature, stresses and strains. This case study will focus on shape, hence modelling since so many other aspects depend on it. In view of the fact that the ultimate representation for models in CAD is solid modelling as it contains all the information required and related to objects such as shape, edges, surfaces and interval volume, this case study focused on modelling the connecting rod using the three most commonly used techniques of boundary representation, solids of revolution and CSG.

Other modelling techniques that are available in most CAD systems include; pure primitive instancing, generalized sweeping (solid of revolution), spatial occupancy enumeration, and cellular decomposition (Du et al., 2018). The three chosen techniques for modelling this connecting rod are the most frequently used because of the wide range of options and different possibilities that they offer to the designer. In particular, CSG was largely employed through the use of several solid primitives that were combined using Boolean algebra, or more commonly referred to as composite modelling.

10.4.1 Modelling the connecting road primitives

The construction and modelling of the connecting rod required multiples of two basic primitives, the cylinder and the box, for a complete CSG process. Table 10.1 shows a summary of the number of primitives required and their location on the connecting rod. As was demonstrated in the previous wheel base assembly, it is important to place the primitives in their correct relative positions before composite modelling. Thus, the construction was derived from just two basic primitives, the cylinder and box, using the appropriate commands as previously explained. Instead of using the default parameter such as length, width and height for the boxes, the coordinate positions of two corners of each box were used. This was particularly useful when it came to locating the primitives in their correct relative positions.

Table 10.1 Connecting road components and primitives

Position on connecting rod	*Primitives required*
Small End	4 cylinders
Big End	6 cylinders
Nuts	4 cylinders
Bolts	6 cylinders
Shank	3 boxes

Although the big-end bearing and its housing were initially modelled as complete cylinders, in real practice, they are split into two. They were deliberately left as such at the modelling stage since most of the components on this end needed to be split. It was therefore more convenient to do it after the Boolean operations using the solid modelling technique of slicing. The only parts of the connecting rod that required a different technique for modelling were the threads for the bolts and nuts.

The threads on the nuts and bolts were a bit cumbersome to obtain using CSG because numerous cylinders would be required to obtain such a profile. Generalized sweeping or solid of revolution was more suitable. The 2-dimensional profile of one-half of the threaded area as shown on the left of Figure 10.14 was revolved about the central axis to obtain the required thread as shown on the right of that figure. The resulting solid thread was then located at the end of the cylinders that form the shank for the bolts. It was also copied and positioned centrally to the nut from which it was subtracted to obtain the internal thread of the nut.

The rest of the components were obtained by using appropriate dimensions and required parameters for the boxes (2 diagonal corners and height) and cylinders (bottom centre, diameter and height. These were then rotated and positioned as shown in Figure 10.15.

10.4.2 Composite modelling of the connecting rod

Boolean operations must be carried out in the correct sequence to avoid any unexpected results. For the small-end bearing housing, the internal cylinder was subtracted from the external cylinder to obtain a hole through which the small-end bearing fits as shown in Figure 10.15. The small-end bearing was obtained in a similar manner. The big end of the connecting rod has more components than the small end, hence more care has to be taken in

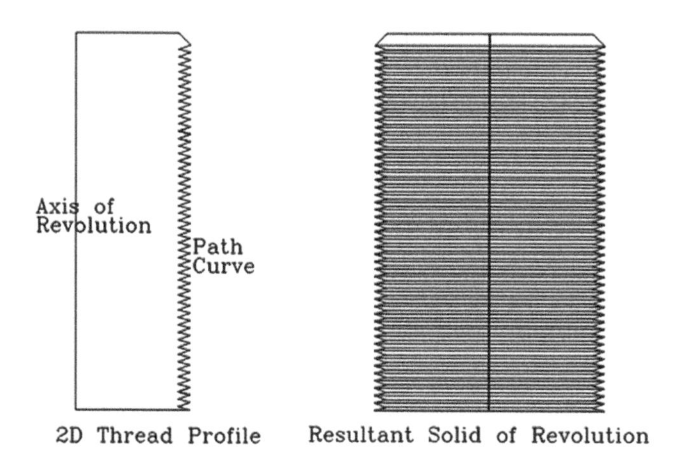

Figure 10.14 Path curve and profile of solid models of the thread.

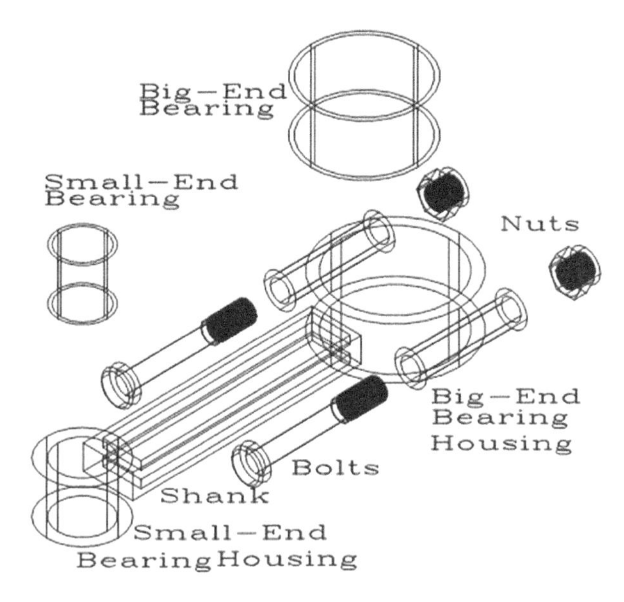

Figure 10.15 Connecting rod primitives in relative positions before boolean operations.

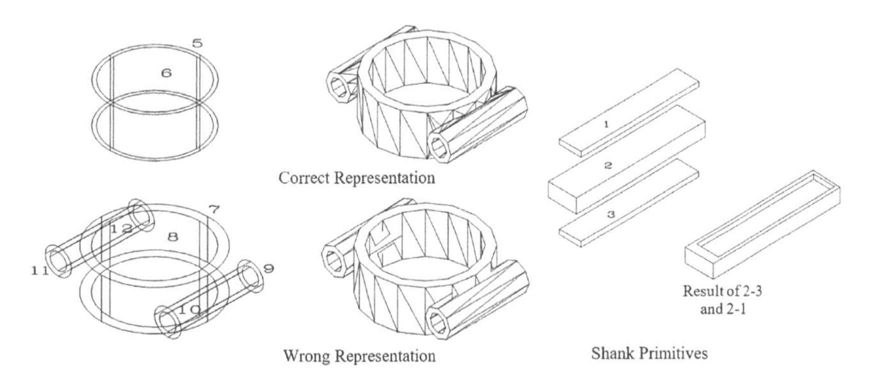

Figure 10.16 Big-end bearing and shank components of the connecting rod.

applying the Boolean operations. Figure 10.16 shows the primitives required to build the big end except for the bolts and nuts. For the big-end bearing housing, the following sequence of operations was carried out; the union of Cylinders 7, 9 and 11 then subtraction of cylinders 8, 10 and 12 from the resulting union (Procedure A). Another alternative that may appear plausible is; cylinder 7 subtract cylinder 8, cylinder 9 subtract cylinder 10, cylinder 11 subtract cylinder 12, and then the union of the results (Procedure B). However, users need to be careful as this second procedure will result in undesirable consequences as shown on the right of Figure 10.16. The inside of the big-end bearing housing obtained by the second procedure appears to have a crack and this is due to the fact that the hole where the big-end bearing fits was created before uniting cylinders 9 and 11. On uniting these two cylinders, a protrusion would obviously result because part of the area through which the cylinders pass, would be hollow.

The shank that connects the bearing housings for the big and small ends was constructed from three boxes as shown in Figure 10.16. Boxes 1 and 3 were first located within box 2 flat at the top and bottom and then subtracted from it to obtain the recess. The shank was then united with the big-end and small-end bearing housings. The special nut for securing the big end as shown in Figure 10.15 was constructed from the union of two cylinders. The hexagonal profile of the nut was obtained by applying chamfers on the four corners of the larger cylinder. The thread generated by a solid of revolution (Figure 10.14) was located centrally and subtracted from the union of the two cylinders to obtain the internal thread. The bolt was constructed from three cylinders that form the head, shank and unthreaded portion. The threaded portion was taken from the solid of revolution (Figure 10.14) and united with the three cylinders and a chamfer was inserted on the head of the bolt to obtain the final model of the bolt. Fillets were applied on those edges between and resulting from the union of the shank with the big and small end bearing housings, with suitable radii. The edges of the boxes forming the recesses on the shank as shown in Figure 10.16 were also filleted to obtain smooth profiles. The inside edges of both ends of the cylinders forming the big-end bearing housing were chamfered even though at this stage, the component parts forming this end are yet to be split.

10.4.3 Assembling the connecting rod

After all the component parts of the connecting rod have been modelled and placed in their correct relative positions (Figure 10.15) and before assembling, it was more convenient to change the colours of each component to visualize the parts easily when moving them from their construction positions to the assembly positions. Alternatively, this could also be accomplished by zooming in and out to avoid the clutter of lines, arcs and circles. The object snap points such as endpoint, midpoint and centre are critical for the assembling process to ensure that the components are fitted into the correct positions.

Firstly, the small-end bearing was moved from its construction position by using the centre of either end as the reference and repositioned on the corresponding side of the small-end bearing housing. Similarly, the big-end bearing was also moved from its construction position to the relative assembly position of the big-end of the connecting rod. The two bolts were moved and relocated to the two holes created on the big end using the corresponding centres of the bolt heads and the centres of the holes. Finally, the nuts were moved from their construction positions to the thread ends of the bolts

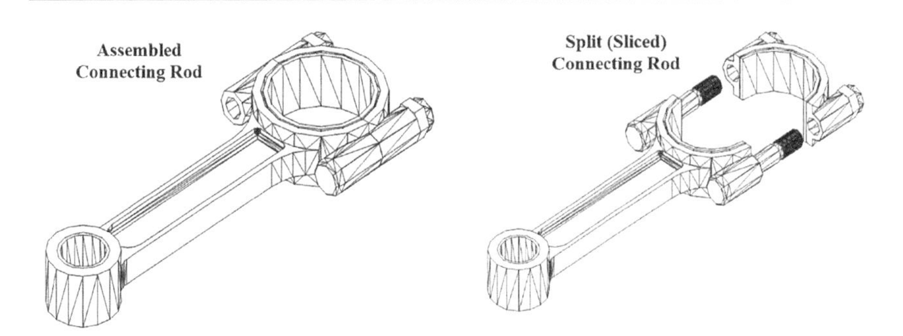

Figure 10.17 Assembled and sliced model of the connecting rod.

(tightening). The resulting model of the connecting rod assembling is shown in Figure 10.17 (left) with hidden lines removed.

10.4.4 Sectioning (slicing) model of assembly

In real practice, the big-end bearing housing and its bearing are split into two to allow for the assembling of the Connecting Rod to the crankshaft of an engine. The set of bolts and nuts is used to secure the split big-end. During the process of modelling the connecting rod, the big-end bearing housing and the bearing were deliberately left as single entities because it was easier and faster to model them using the basic primitives, in this case, cylinders. Most solid modelling systems would have options for cutting or slicing models along defined planes as explained and carried out in the case of the wheel base assembly. The big end was therefore sliced along its central axis by specifying three points on the quadrants of any of the cylinders.

Care must be taken not to slice the bolts as well. This can be accomplished by either removing them temporarily from the assembly or turning off their layers if they were created using layers. Apart from providing the facility for slicing components, selected parts can be moved to new locations or modelled using layers to allow for better visualization of the sliced models. In the case of the Connecting Rod, the nuts and one-half of the big end can be shifted into a new position away from the main model to produce an exploded view such as that illustrated in Figure 10.17 (right).

10.4.5 Other derivatives from the modelled connecting rod

As alluded to earlier, the AutoCAD 3D solid modelling system allows for the automatic conversion of 3D models to 2D orthographic drawings, which are useful for production and construction sites. The assembled model of the connecting rod was converted to 2D orthographic views using the appropriate 3D views option in AutoCAD as shown in Figure 10.18 (left). Only a

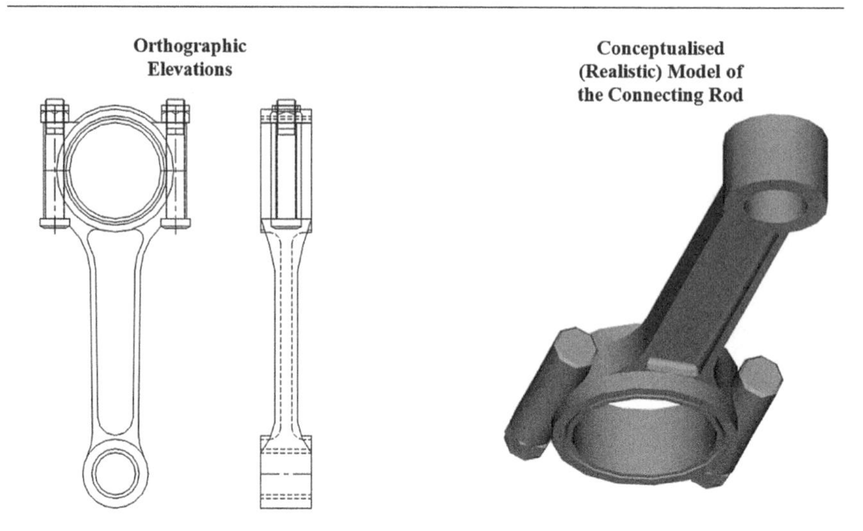

Figure 10.18 Automatically generated orthographic views and conceptualized model.

few adjustments would be necessary to add dimensions and specifications to the 2D orthographic views, thereby bypassing the need for 2D drafting.

While CAD 3D modelling systems have a facility for slicing models as with the case of the connecting rod, this allows designers to carry out internal checks, especially for assembled objects that fit into each other or where some components may be hidden behind others. In most cases, assembled objects have intersecting components and these can be checked to see if there are any shared volumes, and possible errors of interference can be readily identified and corrected.

To make full use of the available facilities in 3D modelling, including analysis, it is advisable to always specify the material being modelled, as was done with the wheel base assembly. With specified materials, 3D models can thus be used to determine engineering mass properties, such as volume, weight, centroids and products and moments of inertia, to ease design changes of the product under development. Part programming information required to produce an object on a Computer Numerically Controlled machine can be generated from the 3D to avoid duplication when Design and Manufacture are integrated. 3D models are also handy for marketing new products through visual displays in brochures.

The ultimate goal in modelling is to produce photo-realistically rendered images that provide clarity of information. Figure 10.18 (right) shows such a rendered image of the connecting rod, derived from the model. Several other visual styles are available to visualize modelled objects as well as the ability to view them from different angles. The 3D geometric model of the connecting rod as shown in Figure 10.17, holds all the information necessary for the automatic generation of a finite element mesh. The mesh can be used for analysing stresses and strains of the connecting rod while in operation, thereby identifying causes of bending while functioning. Finite element analysis can also be used as a tool to predict the behaviour of engineering components, thereby scheduling maintenance ahead of failure. Most CAD solid modelling systems being produced nowadays, such as Solid Works have the capability of carrying out finite element analysis on modelled components.

10.5 SUMMARY

The ultimate aim for CAD systems and 3D modelling, in particular, is the ability and skills required to not only model but analyse components as well. Various skills and techniques were introduced in previous chapters. This chapter culminated in the practical application of these in two examples, a tutorial and a case study. The wheelbase assembly was presented in the form of a step-by-step tutorial of firstly modelling the seven components that make up the assembly, followed by how to manage them through modifications such as rotating and mirroring in 3D and eventually moving the parts into their correct positions of assembly. The second was a case study for modelling a connecting rod for a diesel engine was presented. This was derived from a heavy vehicle maintenance and service company where such engineering parts frequently failed due to motions resulting in bending. This case was however limited to the modelling and assembling of the connecting, which would be useful for input into finite element analysis programs to study the behaviour of the component in operation.

The two practical examples were presented in such a way as to bring out certain salient techniques that should be employed for best practice and correct representation of objects. Such techniques include; the appropriate sequence for applying Boolean operations in CSG to avoid undesirable consequences, accurate derivations of path curves and their positioning to avoid misalignments in assembly, optimal use of the three techniques of B-Rep, CSG and solids of revolution and the appropriate choice of the techniques for efficiency in modelling. The inclusion and specification of materials for each component in the assembly also allows users the ability to draw up a materials property schedule that includes, weight and mass, moments of inertia, centres of gravity, checking for interferences of parts etc., which are necessary parameters in the design of engineering components.

10.6 REVIEW EXERCISES

1. Outline the main CAD 3D modelling operations required in developing the model of an assembly of components in a machine, paying particular attention to duplication of components, alignment, movement and correct positioning for assembly.

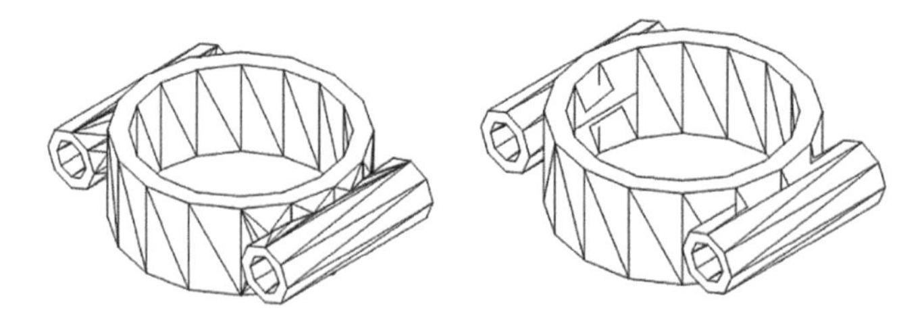

Figure 10.19 3D models of the assembly of the big-end bearing of connecting rod.

2. In order to cut across a machine assembly in CAD, there are various parameters that must be set. List these and explain how they are carried out and the difference between slicing a portion completely off or slicing but retaining it.
3. AutoCAD offers the facility to draw up a materials property schedule from assembled components. The schedule contains several parameters that are useful in engineering design. List and explain at least five of these and their use in engineering design.
4. 'With 3D models, there is no need for 2D orthographic drafting'. Explain whether this is a true statement or not, outlining the pros and cons of straight 2D and drafting and that of automatically generating 2D orthographic projections from 3D models.
5. Figure 10.19 shows two models of the assembly of the big-end bearing of a connecting rod in hidden visual style. The one on the right clearly shows a crack within it. Explain how this could have arisen during 3D modelling and assembling of the connecting rod and how this can be resolved.

Chapter 11

Customization of CAD software

11.1 INTRODUCTION

Just like many CAD software packages, AutoCAD is a general and multi-purpose drafting and modelling software that can be used by many disciplines, such as electrical, mechanical, civil and structural engineering, architecture, mining and metallurgical engineering, cadastral surveying, cartography and geoinformatics, etc. A package of such diverse nature could only be supplied in a general format such that anyone in the various fields wishing to use it can have easy access. As such it is provided in a very flexible and wide manner to enable the fast and efficient production of working drawings and models.

However, this flexibility invariably creates a challenge when a user is faced with so many possibilities which may result in delays in the process of navigating through or developing required specific presentations. Users in the various fields need to have the ability to control how AutoCAD operates or behaves by customizing a number of facilities that may be peculiar to their own needs. As a consequence of this multitude of possibilities, most CAD software companies have developed derivatives of their main packages to suit particular applications such as AutoCAD Architecture, specifically for architecture, AutoCAD Plant and Instrumentation Design, etc.

In terms of affordability, they have further developed lighter versions of their software to enable access to those who would not necessarily require the full systems, hence the LT versions of the various releases of the AutoCAD software. In addition, the general-purpose AutoCAD software is also supplied with capabilities and options to adjust it to suit specific disciplines and applications. These adjustments include the ability to add one's own customized icon and pull-down menus, command aliases as well as customized line types and hatch patterns. Autodesk supplies AutoCAD with a number of support files that are written in the American Standard Code for Information Interchange (ASCII) code and these files can be accessed and modified using standard text editors. In this chapter, various ways in which these ASCII files can be manipulated and adjusted in order to make AutoCAD suit our

DOI: 10.1201/9781003288626-11

specific needs will be discussed. This will not only result in a package that is user-friendly and easy to use but also increases productivity, otherwise without such facilities, the use of CAD remains a fallacy to some who may not be adequately skilled to use it to its full potential and derive the many benefits.

This chapter will therefore cover, ways in which users can modify AutoCAD program parameters, the use of command aliases, pull-down and screen menus, adding of icon menus, use of script files and customizing line types and hatch patterns. These guidelines will provide ways how to customize AutoCAD to specific needs, aimed at the efficient production and development of engineering drawings and models, further justifying the need to invest in CAD systems for the ever-growing and competitive global economies. A case y for a power and electricity and distribution company, where they were assisted to customize their menus for the efficient production of drawings and models for their power stations, is also included to demonstrate the practical application of customization in everyday practice. This adjustment coincided with the company's migration from manual to CAD drafting, hence it could not have come at a better time to justify their investment in CAD.

11.2 PROGRAM PARAMETERS

Program parameters in AutoCAD allow users to enter and operate external commands from within AutoCAD. The instructions to execute such are contained in the **acad.pgp** which is located in the AutoCAD support directory. This is a typical text file that can be accessed and edited using the AutoCAD command, AI_EDITCUSTFILE for the later versions of AutoCAD or edit acad.pgp from the DOS command of the older versions of AutoCAD. For the later versions of AutoCAD, the user is prompted for a file name, of which **acad.pgp** is entered. Either way, the text screen with the editable ASCII file will be brought up. This is the file where command aliases such as L for line, C for circle etc. are defined. If the PGP file is edited while the program is running, it is necessary to reinitialize it by entering REINIT to use the revised file or restarting AutoCAD to reload the file automatically. There are five key parameters on each line of the PGP file, which would ordinarily appear as follows:

CATALOG, DIR /W,0, File Specification: ,0
DIR, DIR, 0, File Specification:, 0
EDIT, EDLIN, 0, File to edit:, 4
SHELL,, 0,*OS Command:, 4

The structure of each line contains the following attributes: Command Name, OS Request, Memory Reserve, Prompt:, Return Code. These attributes have the following meanings:

Command Name:	This is the command that is typed and entered at the AutoCAD Command prompt in order to execute that particular command.
OS Request:	The command string that is sent to the operating system in order for the command name (above) to be executed.
Memory Reserve:	This attribute serves to maintain compatibility with previous versions of AutoCAD. This value is rarely used and hence is normally set to 0.
Prompt:	After entering the command name, a user may wish to let AutoCAD prompt the user what the next step is. This is where you fill in the set of words asking the user to do something. If no prompt is required, it can be left blank with a comma at the end.
Return Code:	The last attribute is a bit or number 0, 1, 2 or 4 which controls the way AutoCAD behaves after the execution of the command is complete. The meanings of the numbers are as follows: 0: Return to AutoCAD Text Screen 1: Load a named DXB file 2: Construct a block from a name DXB file 4: Return to the Text or Graphics mode that was current when the command was started.

11.2.1 Modifying the program parameters

Having understood the structure of the ASCII **acad.pgp** file, the next step would be to make some changes or modifications to it by putting in some commands that will allow the editing of ASCII text files from within AutoCAD. This can be done by changing the first line of the acad.pgp file from **EDIT, EDLIN, 0, File to edit:,0** to **EDIT, EDIT, 0, File to Modify:, 4**.

After modifying the file, it is necessary to save it in ASCII format by using the normal File and then Save it within the Notepad or text editor. However, for this to be effected, the acad.pgp must be in the AutoCAD support directory and should be reinitialized or AutoCAD should be started afresh.

The changes can be tested by typing **edit** and enter at the AutoCAD command prompt, followed by typing in the **acad.pgp** file in a similar way to the later versions of AutoCAD and this should take the user straight to the editable ASCII file in Notepad. This also means that all future text files including AutoLISP files can now be edited within AutoCAD. The later versions of AutoCAD such as AutoCAD 2021 have been greatly simplified by allowing commands such as AI_EDITCUSTFILE to automatically take the user to the text editor or Notepad by simply specifying the file that needs to be edited. However, it must also be noted that this all depends on the version that is being used. For example, with AutoCAD LT versions, the program parameters file will accordingly be **acadlt.pgp**.

11.2.2 Reinitializing program parameters

If any changes have been made to the **acad.pgp** file, there is really no need to close and restart AutoCAD in order to effect changes but instead, the

REINIT command can be used to quicken the process. For example, to add a command that will list all the drawings in the current directory, the following steps are taken:

1. The **acad.pgp** file is opened using AI_EDITCUSTFILE .
2. Immediately after the last line on 'Examples of External Commands' the following line can be added:

   dwgs, dir/p *.dwg,0,0

3. The changes are then saved to return to the AutoCAD graphics screen.
4. In order for the above change to be effected, the system is reinitialized by REINIT and a dialogue box will pop up with a number of options. Clicking inside the box alongside the words PGP File and followed by OK will update the PGP file.
5. Finally, the command, DWGS at the command prompt will bring up and list all drawings in the current directory.

11.3 COMMAND ALIASES

To quicken the process of instructing AutoCAD to execute a particular command, many of the commands have been abbreviated to command aliases or shortcuts to avoid wasting time typing the full command. These include: L for line, C for circle, E for erase, M for move, MI for mirror, etc. The AutoCAD packages are provided with these having been defined and are at the disposal of the user. However, users have the ability to further define additional command aliases by customizing the **acad.pgp** file to suit their requirements, as long as they do not clash with those that are provided by AutoCAD as defaults. Command aliases are also part of the editable PGP file which is written in the ASCII code and can be adjusted or added to the AutoCAD system similar to the general external commands covered in the program parameters section above. The structure for this section on command aliases is as follows:

Alias, *Command Name

Alias:	This is the abbreviation or command that a user may wish to apply in AutoCAD.
Command Name:	This is the normal AutoCAD Command written in full and preceded by an asterisk (*).

For example, in order to shorten long commands such as AI_EDITCUSTFILE, the one used to edit the PGP file, the acad.pgp file is opened and the user can scroll down to find the particular command, which will be in alphabetical order, then add or adjust the command, AI_EDITCUSTFILE as follows:

```
PGPEDIT, * AI_EDITCUSTFILE
```

This is followed by saving the PGP file and then reinitializing AutoCAD using REINIT so that in future the user does not need to type in the full command AI_EDITCUSTFILE but instead type in the short form, PGPEDIT. Users must ensure that appropriate shortcuts for AutoCAD commands are selected, otherwise it may clash with existing ones. This way of quickening the entering of commands is only limited by the number of available abbreviations that can be used. Companies wishing to adopt this have to standardize their command aliases to ensure that all departments making use of the same CAD system are familiar with the adopted command aliases.

11.4 CUSTOMIZING MENUS

The Slide and Pull-Down menus, as shown in Figure 1.7, that are displayed on the AutoCAD screen at start up are the default menus that are supplied with AutoCAD. However, depending on a user's discipline, they may wish to find some of the menus unnecessary or better still they might wish to adjust or add their own menus alongside those supplied by AutoCAD. This will greatly simplify the use of AutoCAD as will be demonstrated in the case study.

The appearance of the two menus is programmed using the ASCII code as in the program parameters and command aliases covered above. These adjustments can be made in the AutoCAD menu file, **acad.mnu** which is also located in the AutoCAD support directory. However, extra care has to be taken to program or make changes to this file, otherwise it may create some complications. One way to get over this would be to use the tutorial menu, **tutor.mnu** to experiment first before making changes to the main AutoCAD menu file, acad.mnu. The tutorial menu file can be accessed in the AutoCAD Tutorial directory and is also an editable ASCII file.

11.4.1 Menu file structure (earlier versions)

The following guidelines are applicable to earlier versions of AutoCAD up to Release 12. The newer versions, including AutoCAD 2021 on which this book is mostly based are found towards the end of this section. The tutorial menu can be accessed by typing the EDIT command and then specifying the tutorial path to edit the ASCII file **tutor.mnu**. Both the tutorial menu and the main AutoCAD menu have the same general structure. The names for each of the sections in these ASCII files are preceded by three asterisks (***) and the seven sections are as follows:

*****BUTTONS**:	Allows programming of the different buttons of a pointing device. However, the first button is always the pick button and cannot be reassigned.

***AUX:	Allows for the configuration of the system mouse buttons and is used in the same way as the BUTTONS.
***POP*n*:	Creates the Pull-Down menu bars at the top of the AutoCAD screen when the mouse is moved into that area. The **POP0** menu is the cursor or object snap menu obtained by pressing Shift together with the RMB.
***ICON:	Allows for the insertion of menus displaying list boxes and slides (images). Useful for displaying particular blocks in the system for easy picking and insertion of the particular slide into the drawing.
***SCREEN:	Controls screen menus found on the right-hand side of the AutoCAD screen. Supplements tablet in the event that the system has no tablet.
***TABLET:	Similar to the Screen and can also be customized to particular needs.
***COMMENT:	Allows users to add comments in their menu files.
	This section will deal with the POP*n*, ICON and SCREEN Menus since these are the ones commonly necessary to modify from the default AutoCAD settings.

11.4.2 Pull-down menus

The Pull-Down Menus on the AutoCAD screen start from File, Edit, View, Insert through to Parametric, Window and then Express. In the ASCII menu files, these are referred to as POP*n* menus where *n* is the position of the Pull-Down Menu on the screen. For example, the POP1 menu is the File Pull-Down menu, POP2 is the Edit Pull-Down menu etc. The POP0 menu as explained above is the Cursor Object Snap menu obtained by pressing Shift and RMB. The general structure of the POP menus can best be explained by using one of them, for example, POP1, with only a few of the pull-down items displayed to explain the different parts. These are the default AutoCAD settings. The expression or phrase in [square brackets] is what will be displayed on the screen.

*****POP1**	Definition of the POP menu preceded by (***)
[File]	First item in the POP1 menu is File
[New...]^C^C_new	Sub-item 'New' under the 'File' heading
[Open...]^C^C_open	Sub-item 'Open' under the 'File' heading
[Save...]^C^C_qsave	Sub-item 'Save' under the 'File' heading
[Save As...]^C^C_saveas	Sub-item 'Save As' under the 'File' heading
[Recover...]^C^C_recover	Sub-item 'Recover' under the 'File' heading
[--]	Allows for the demarcation between sections of the menu
[Plot...]^C^C_plot	Sub-item 'Plot' under the 'File' heading
[--]	Creates a demarcation line between menu items
[Exit AutoCAD]^C^C_quit	Sub-item 'Exit AutoCAD' under the 'File' heading

After each sub-menu, there is ^C^C_ and then an AutoCAD command. ^C^C (Ctrl-C) is used to terminate any other command that may be in operation before the required command is activated. Normally ^C would be sufficient to do this but there are some AutoCAD commands that require to be terminated by pressing Ctrl-C twice, for example, dimensions. The command that comes after the sub-item is preceded by the underscore (_). To change any of the headings that appear on the screen, the text editor or Notepad can be used to load the **tutor.mnu** in the tutorial directory (Release 12 and below). For instance, under the POP5 menu, the title **Modify** can be changed to **Change**.

In general, using the information and description for the POP menus, the words or phrases in [square brackets] can be changed to the user's preference. The edited menu needs to be saved and then from AutoCAD, the command MENU prompts the user to specify which menu must be loaded from the appropriate directory. For AutoCAD 2021, the AutoCAD menu will be located in the support directory but as **acad.cuix**. This will be dealt with in more detail in the later sections of this chapter. Loading the revised menu will automatically effect all the changes that would have been made.

11.4.3 Cascading pull-down menus

Some of the sub-items in the Pull-Down menus also have sub-items under them. These are referred to as cascading menus. For example, if a menu item **Attribute** is to be added as the last item on the POP5 menu, which was edited from Modify to Change in the last section, the **tutor.mnu** can be opened in a text editor to make the adjustments. Assuming that the sub-items to be added to this POP5 menu under **Attribute** are: **Definition, Display Control** and **Control** *Value* in a cascading menu, then the following changes will be made to the POP5 menu. The bold text at the bottom of the POP5 menu section can be added as explained alongside the added phrases.

***POP5**	Definition of the POP menu preceded by (***)
[Change]	Title of the POP5 Menu is 'Change'
[Erase]^C^C_erase	Sub-item 'Erase' under the 'Change' heading
[Extend]^C^C_extend	Sub-item 'Extend' under the 'Change' heading
[Trim]^C^C_trim	Sub-item 'Trim' under the 'Change' heading
[--]	Creates a demarcation line between menu items
[Move]^C^C_move	Sub-item 'Move' under the 'Change' heading
[Rotate]^C^C_rotate	Sub-item 'Rotate' under the 'Change' heading
[Scale]^C^C_scale	Sub-item 'Scale' under the 'Change' heading
[Stretch]^C^C_stretch crossing	Sub-item 'Stretch' under the 'Change' heading
[--]	Creates a demarcation line between menu items
[->Attribute...]	Menu item 'Attribute' as heading for the Cascade.

[Definition...]	Cascaded menu item, 'Definition' under 'Attribute'.
[Display Control]	| Cascaded menu item, 'Display Control' under Attribute.|
[Control Value]	Cascaded menu item, 'Control Value' under 'Attribute'

Commands can be added after the cascaded menu items. This sign (≥) is placed before a heading to indicate that items after this will be cascaded under that heading. The text file can then be saved and reloaded (recompiled) in AutoCAD using the MENU command.

11.4.4 On-screen menus

The On-Screen menu that appears on the right-hand side of the AutoCAD screen can be modified in the same manner as explained above. The structure of the first section of the ***SCREEN is as follows:

*****SCREEN**	Display on Screen
****S**	
[AutoCAD]^C^C^P(ai_rootmenus) ^P	AutoCAD
[* * * *]$S=OSNAPB	* * * *
[BLOCKS]$S=X $S=BL	BLOCKS
[DIM:]^C^C_DIM	DIM
[DISPLAY]$S=X $S=DS	DISPLAY
[DRAW]$S=X $S=DR	DRAW
[EDIT]$S=X $S=ED	EDIT
[INQUIRY]$S=X $S=INQ	INQUIRY
[LAYER...]^C^C_DDLMODES	LAYER...
[MVIEW]$S=MVIEW	MVIEW
[PLOT...]^C^C_PLOT	PLOT...
[SETTINGS]$S=X $S=SET	SETTINGS
[UCS:]^C^C_UCS	UCS:
[UTILITY]$S=X $S=UT	UTILITY
[SAVE:]^C^C_QSAVE	SAVE

11.5 CUSTOMIZED PULL-DOWN MENUS

Users have the flexibility to adjust the appearance of their AutoCAD screens to suit their needs and applications. Some of the pull-down menus that are supplied by AutoCAD as standard defaults may not be necessary for some organizations. At the same time, some organizations may want to add their own pull-down menus to enhance their operations in the efficient production of models and drawings as will be demonstrated in the case study for a power and electricity distribution company. For the benefit of AutoCAD users who are still operating the earlier versions of the package, this section will be split into 2, firstly on how to customize and add pull-down menus in earlier versions of AutoCAD such as AutoCAD Release 12 and secondly, how to do the same in later versions such as AutoCAD 2021 on which this book is mainly based. This is in line with one of the objectives for this book in that, where necessary, demonstrations will be made for areas where Autodesk has made noticeable changes to the software, hence the need to clarify these for a smooth transition from one version to another.

11.5.1 Customizing pull-down menus – earlier versions

The most significant advantage to customize CAD systems is the ability to add user-friendly pull-down menus that are familiar or peculiar to an organization or remove some of the default menus supplied with AutoCAD that are rarely or never used. To remove menus, users need only delete the corresponding POP menu from the ASCII menu file and then recompile it.

However, this should only be done once an organization has made a decision to adopt a new structure. Otherwise, unilaterally changing the settings may actually create more problems than desirable. The changes must be done for continuity in such a way that a new design engineer should find the changes useful to access for the speedy production of drawings and models. It is also advisable to carry out these modifications in the Tutor Menu before applying them to the actual AutoCAD menu. Alternatively, first, save the **tutor.mnu** file to **tutorold.mnu**. This is useful in case a user makes a mistake that they are not able to trace. To demonstrate the application, a new POP menu will be added. Depending on which POP menu is available in the version being used, the next number will be used for the new POP menu. For instance, for those with versions ending at POP7, then the next will be POP8 to be labelled as Architecture and coded using a suitable text editor to put an additional pull-down menu, Architecture with sub-items, Windows, Doors and Electricals as follows:

*****POP8**
[Architecture]
[Windows]$i=fronticon1 $i=*
[Doors]$i=fronticon2 $i=*
[Electricals]$i=fronticon3 $I=*

The **$i** sign is for displaying a named icon menu such as fronticon1, fronticon2 and fronticon3. These are defined in the ***ICON menu section which can be located by scrolling further down the menu file. Leaving a blank line after the last line in the ***ICON section, the three items above will then be added and defined as follows:

****fronticon1**
[Windows]
****fronticon2**

[Doors]
**fronticon3
[Electricals]

The menu file is then saved in the text editor and recompiled in AutoCAD using the MENU command. The additional pull-down menu, Architecture will thus appear as the last pull-down menu on the AutoCAD screen. Clicking on this new menu will display the three sub-items under it, that is, Windows, Doors and Electricals. Further clicking on any one of the sub-items, a blank icon menu box appears with the name of that sub-item. Naturally, the next step will therefore be to look at how to fill in the icon menu box with pictures of, for example, blocks that may need to be inserted into a drawing. This will be dealt with in Section 11.6.

11.5.2 Customizing menus and toolbars – AutoCAD 2021

While the previous sections in this chapter focused on how menus and toolbars were programmed and modified in the earlier versions of AutoCAD, this section will show the migration to the later versions of the package. In most of the newer versions of AutoCAD, Autodesk has simplified customization of menus and toolbars to the extent that it is user-friendly and with no need to hassle for opening, editing and saving menu and program parameters files in some text editors outside AutoCAD before reloading or recompiling for the changes to take effect. This is one of the reasons the newer versions such as AutoCAD 2021 do not have accessible or editable menu files anymore, such as acad.mnu, acad.mns, acad.mnc, etc. The menu file for AutoCAD 2021 is now stored in a non-editable version, **acad.cuix**.

The customization files can now be accessed in a simpler manner through the CUI command which leads to the Customization User Interface dialogue box shown in Figure 11.1. The customization dialogue box basically consists of four sections: the customization workspaces such as toolbars, menus, ribbon etc., the command list associated with any particular item in the first section that would have been selected to be customized, the properties associated with the customization option selected and general.

Apart from congested or too many unnecessary menus and sub-menus, there could also be congested toolbars containing icons or commands that a user may never make use of. AutoCAD offers a facility where users can adjust the number of icons within a toolbar or even create their own containing icons that they normally use, depending on their specialization. Using the dialogue box in Figure 11.1, customized toolbars can be created by first selecting where the user wishes to place them.

To demonstrate this, and assuming that a new toolbar, **My Custom Bar** is required to be added, the **Toolbar** in the first section of the dialogue box is selected and right clicked, then **New Toolbar** and the name **My Custom Bar** can be typed in. Once the new toolbar has been added, the next thing would be to add to it, the commands that a user wishes to associate this toolbar

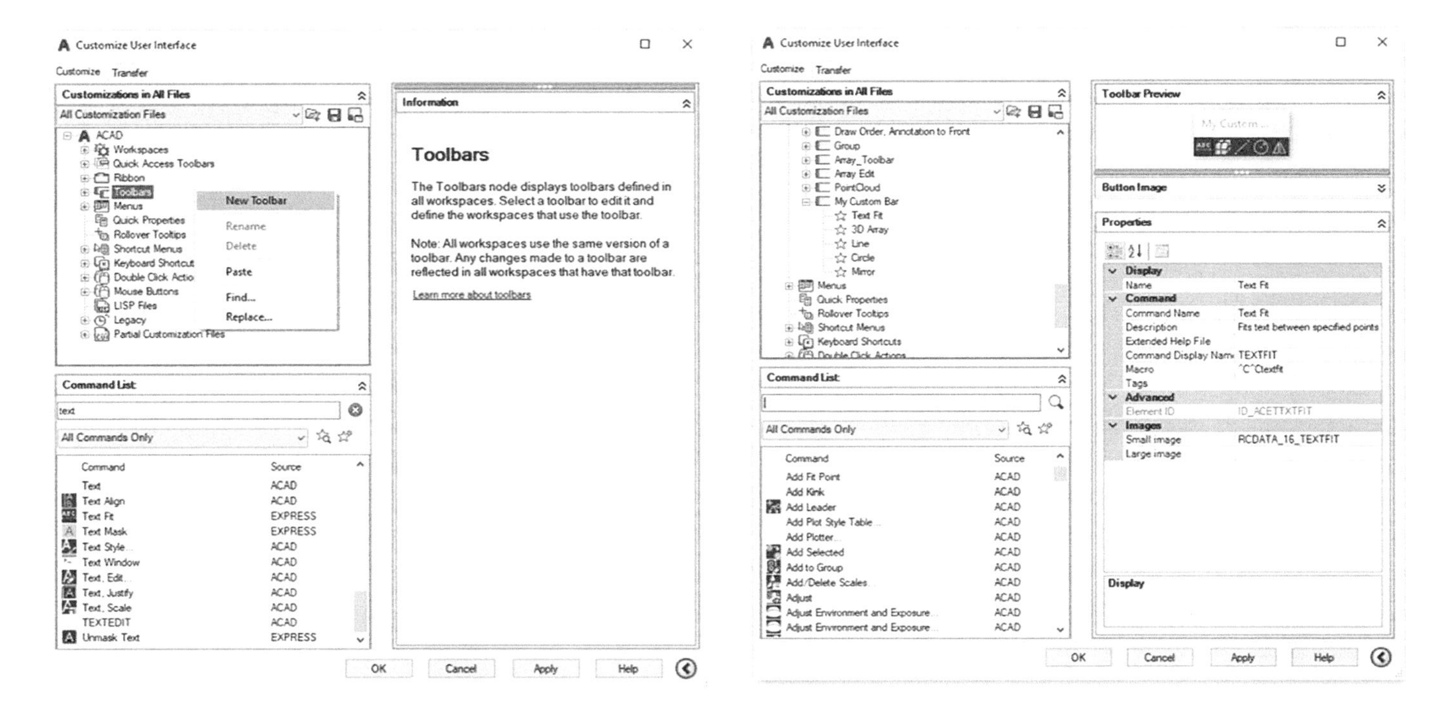

Figure 11.1 Customization user interface dialogue box.

with, or those command icons that should fall under the new toolbar. All the AutoCAD commands are contained within the Command List just below the **Customization in All Files** section as shown in Figure 11.1.

Assuming that the user frequently uses the commands, Text Fit, 3D Array, Line, Circle and Mirror, these are then selected one at a time and dragged to the new toolbar or by right-clicking on each, copied and then pasted in the new toolbar, to eventually look as shown on the right of Figure 11.1. Because of the numerous commands available in AutoCAD, it might be time-consuming to scroll through all of them in order to come to the desirable command before dragging it to the new toolbar. AutoCAD commands in this section are arranged in alphabetical order. However, some have duplication of the same words or phrases for different commands. In this case, the user can type in a keyword just above the command list and all commands associated with that keyword will be shown from where the desirable command can be selected and dragged to the new toolbar.

The five items, Line, Circle, Text, Mirror and 3D array that have been added to the new toolbar, **My Custom Bar**, can be reordered by moving them around or simply maintaining the order in which they were added. At the top right of the Customization User Interface dialogue, a preview of the new toolbar will also appear with the new name and icons for the selected AutoCAD commands. This effectively completes the creation of a custom toolbar. However, before applying the changes, it is important to make sure that the selected commands for the new toolbar appear in whatever workspace the user is logged on. In this case, there is a need to scroll up and select **Workspaces** under **All Customization Files** on the dialogue box then select the current or appropriate workspace that is being used. In the top right corner of the dialogue box, the contents for that particular workspace will be displayed, together with the new toolbar, **My Custom Bar**. This can be further customized by clicking on customize, and additional command icons can be selected to be added to the current workspace customization. Once satisfied, the Apply button on the dialogue box can be clicked, followed by OK and the new toolbar, **My Custom Bar** will be dumped on the graphics screen from where it can be dragged and docked alongside other toolbars.

The creation of customized menus is done in the same manner as selecting Menu from All Customization Files on the dialogue box and then right-clicking it and New Menu. Assuming here that a new pull-down menu, Architecture is required, this can be typed over the Menu1 name. This will automatically and by default be placed immediately after the last pull-down menu, Express. However, it can be moved around by dragging it to any particular position of choice within the All Customization Files section of the dialogue box. In the same manner, as covered under toolbars, desirable commands can be selected from the command list and dragged to the new menu, Architecture. As an example for this one, five AutoCAD command icons, Multiline, Ellipse, Hatch, 3D Rotate and Array are dragged and added to the new menu, Architecture as shown in Figure 11.2 (left). Again, these can

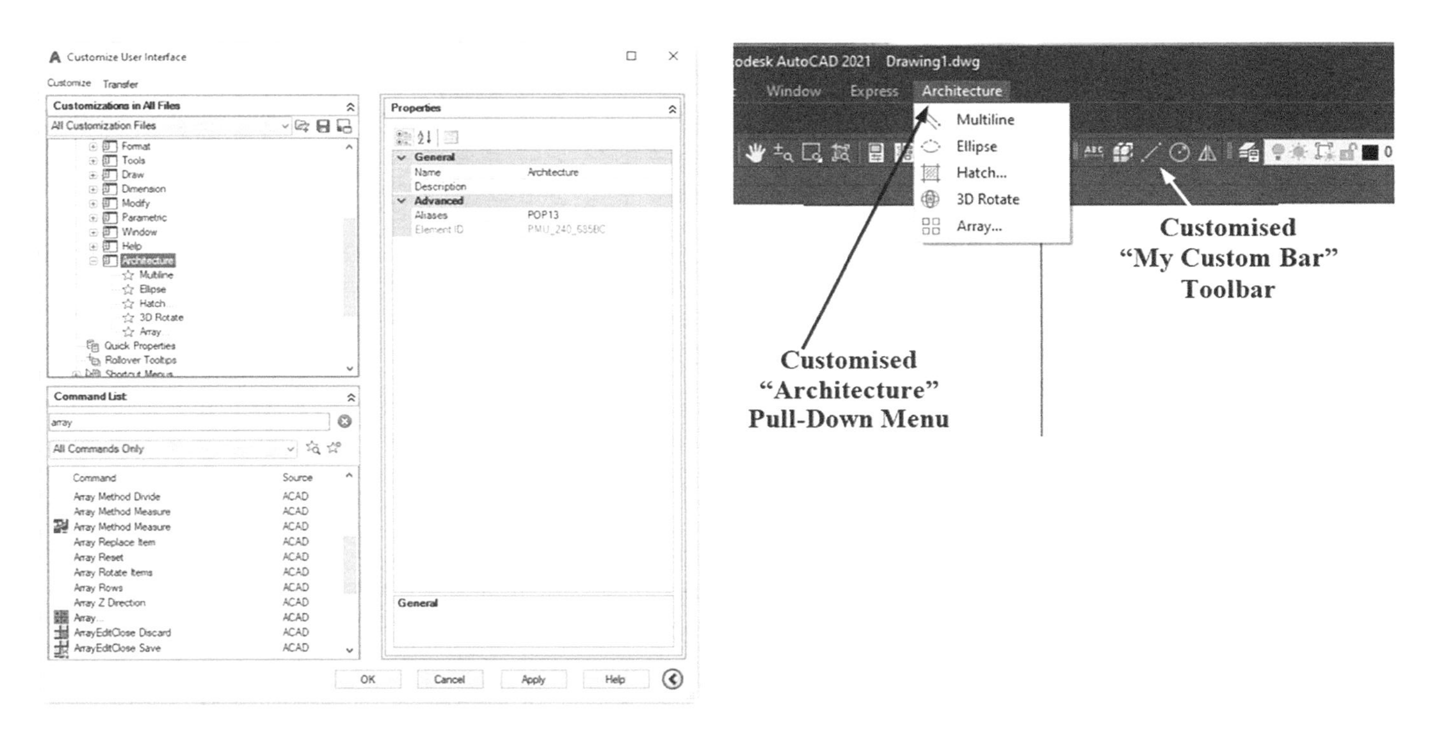

Figure 11.2 Customized menus.

be rearranged by moving them around in any desirable order. The changes are then effected by Apply and OK. Automatically, the pull-down menu is loaded and is now part of the AutoCAD menus as shown in Figure 11.2 (right). In the same manner, undesirable command icons can be removed. This is demonstrably easier to customize menus and toolbars using the customize command, CUI as compared to programming menu files in ASCII code in the earlier versions of the software.

11.6 ICON MENUS

All the commands that appear on the Command list of the Customization User Interface are derived from the AutoCAD database of commands. Each one of them has an associated icon such as a line in the case of the command icon LINE, a circle in the case of the command icon CIRCLE etc. AutoCAD offers the facility for users to create their own icons for easy retrieval and usage. The toolbars and pull-down menus that have been created can be further modified to include customized icons with pictures of desirable items such as blocks or symbols for engineering components as covered in Chapter 7. For the Architecture pull-down menu that has been created, symbols or blocks such as doors, windows etc. can be added. The pictures that would have been created as blocks can also be accessed and pasted to become icons that can be accessed by simply clicking on them rather than going through the process of typing the command and locating the block before inserting it into the drawing. However, it is really a matter of preference as some users may find it easier and faster to type in command rather than to search for an image icon on the menu or toolbar.

Typical symbols such as single or double doors, windows, electronic circuit symbols etc. that were created in Chapter 7 can be used to illustrate and demonstrate the use of image icons to be attached to menus and toolbars. Once the blocks have been created, they should be stored in an appropriate and standard directory within AutoCAD, such as **.......\AutoCAD 2021\Architecture Blocks**, that an organization can standardize for ease of access by its employees. At this point, two architectural symbols or blocks, **Single Door.dwg** and **Double Door.dwg** as shown in Figure 11.3 will be used. These can individually be created by lines and arcs as shown in the figure and then saved or written as blocks using the WBLOCK command to be able to access them in any other drawing, ensuring that the base point (insertion point) and the objects that constitute the block are properly selected and saved in a suitable directory such as ....\AutoCAD 2021\Architecture Blocks\Single Door.dwg.

The two examples of architectural symbols, Single Door and Double Door, are the ones that will be used to create pictures to fill in the blank icon menu boxes for customized menus. However, before this can be done, the

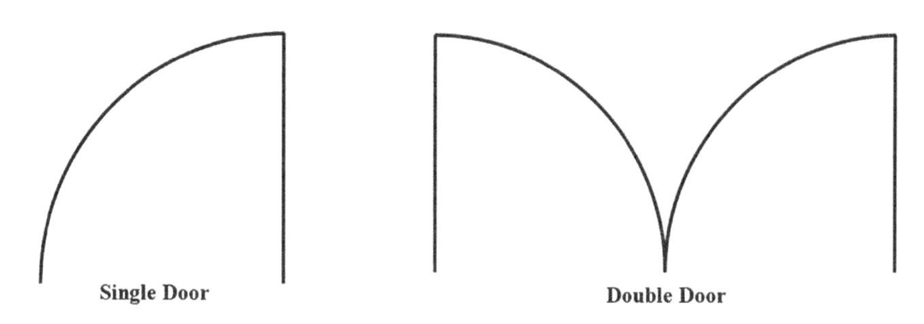

Figure 11.3 Architectural symbols (blocks) to use as slides.

drawings for the two blocks should be converted to slides first in order for them to appear as pictures in the boxes. The guidelines that follow are with reference to the earlier versions where the acad.mnu can be loaded into a suitable text editor to modify the ASCII file in the same manner as described previously. To create the required slides, the two blocks for Single Door and Double door, can be opened individually in AutoCAD from the directory where they were stored. The AutoCAD command makes slide (MSLIDE) entered to bring up the dialogue box with the file name of the particular block opened such as **Single Door.dwg** or **Double Door.dwg**. This will automatically be in the same directory that the blocks would have been created, hence clicking ok will then create the slide file, **Single Door.sld** and **Double Door.sld**. The Tutorial Menu (tutor.mnu) can be used to enter the slides that have just been created so that the blank icon boxes created in Section 11.5.1 can now be filled with pictures of the blocks. A suitable text editor can be used to open the Tutorial Menu, tutor.mnu. Once the tutorial menu is opened, the user can scroll down to the POP8 section for Architecture where the last statements, **fronticon1, **fronticon2 and **fronticon3 were typed. The following lines can then be added so that the revised **fronticon2 of the POP8 section appears as follows:

fronticon2
[Doors]
[Single Door,Single Door]^C^Cinsert Single Door
[Double Door,Double Door]^C^Cinsert Double Door

The ASCII Tutorial Menu file is then saved and the text editor is closed to return to AutoCAD. At the command prompt, the command MENU is used to reload or recompile the Tutorial Menu so that the changes will be effected. The two slides will now appear in the icon boxes. To insert these blocks into any drawing, the Architecture pull-down menu is selected, followed by double clicking the desired icon to insert it in the same manner as inserting a block from a database of blocks.

11.7 AUTOMATION OF TASKS – SCRIPT FILES

AutoCAD can be customized to read a series of instructions from an ASCII file commonly referred to as script files. This is the equivalent of batch files in MS-DOS. Script files can be loaded while working on a drawing and they are usually used to automate tasks and avoid the routine step-by-step of answering questions and entering the required variables before a task is accomplished. The common applications of script files are:

1. Altering settings or display of the current drawing.
2. Displaying a series of slides continuously equivalent to a real slide show.
3. Setting up of configurations for limits, units snap and grid.
4. Adding predetermined layers in a drawing.
5. Ending one drawing and starting another.

All in all, script files help to shorten some of the routine tasks that are normally carried out when constructing a drawing. Instead of waiting to answer one question at a time until something is done, all this can be incorporated into one operation if the answers to the questions are standard, predetermined or known.

This can be illustrated by an example script file which can be created in a suitable text editor, for example, .scr, of which the following text can be entered.

EG

```
; The blank line above acts as Enter and the semi-colon indicates a comment.
filedia 0
menu tutor
line 1,1 3,3

; The blank line above acts as Enter. Only leave one line.
; This completes the line command.
zoom e
```

After saving the script file, example.scr in an appropriate directory, the following can be entered on the DOS prompt ....AutoCAD 2021\Script\ newdwg example. This will start AutoCAD with a new drawing called newdwg and then it will go straight into starting the example script by loading the Tutor Menu, drawing a line from 1,1 to 3,3 and zooming to extents. The semi-colon(;) in the script file indicates a comment and a blank line acts as Enter (↵). Setting filedia to 0 simply turns off the dialogue boxes in AutoCAD. The script command can be used at the AutoCAD Command prompt to start any script files.

As a second example, a script file will be created to set tilemode to 1, setup 3 viewports with an isometric plane in the third viewport and turn on the dialogue boxes. This can be achieved by coding the following script in a suitable text editor.

```
; Example2.scr
; Script file to create 3 viewports, one with the isometric plane.
filedia 0
tilemode 1
viewports 3 r
vpoint 1,1,1
filedia 1
```

After saving the example2.scr and exiting the text editor, the command SCRIPT can be entered in AutoCAD and when prompted for script file, example2 is entered, which results in the creation of three viewports, one with an isometric plane on the graphics screen. Frequently, users are asked by the system, 'Are you sure?'. This can be time-consuming, especially where the answer is obvious but it is meant to make sure that it is not being done by an accidental mistake. However, this can be suppressed by using suitably coded script files through the command EXPERT.

EXPERT is a system variable that controls the issuance of certain 'Are you sure?' prompts, as described below. The value stored is an integer.

Options	
0	(Default) Issues all prompts normally.
1	Suppresses 'About to regen, proceed?' and 'Really want to turn the current layer off?'
2	Suppresses the prompts in Option 1, as well as Block's 'Block already defined. Redefine it?' and Save/Wblocks 'A drawing with this name already exists. Overwrite it?'
3	Suppresses the prompts in Options 1 and 2, as well as those issued by LINETYPE if a linetype is already loaded or creates a new linetype in a file that already defines it.
4	Suppresses the prompts in Options 1–3, as well as those issued by UCS, Save and VPORTS Save if the name supplied already exists.
5	Suppresses the prompts in Options 1–4, as well as those issued by DIM, Save and DIM Override if the dimension style name supplied already exists (the entries are redefined).

When a prompt is suppressed by EXPERT, the operation in question is performed as though the user had responded 'Yes' to the prompt. As an example, users can add in the script file example2.scr, just before filedia 0, the following text, **expert 5**. This will suppress any 'Are you sure?' questions that may be asked in the series of commands that make up a script file. A series of slides can also be viewed using the script command. Blocks such as those created in Chapter 7 or the two used in this chapter (Single and Double Door) can be used to demonstrate the use of scripts in this regard.

The following text can be typed into a suitable text editor and saved as example3.scr

```
; Example3.scr
; Script file to display a series of slides.
expert 5
filedia 0
vslide Single Door
vslide Double Door
filedia 1
```

After saving and exiting the text editor, the script command can be entered in AutoCAD followed by selecting example3. The slides will then be displayed on the screen one after another. The slides can also be displayed continuously using the RSCRIPT command at the end of the script file and also specifying a delay, that is, time to pause between slides. The time is specified in milliseconds. Script Example3 can be adjusted to put the RSCRIPT command and appropriate delays as follows, to display the two slides continuously with a pause of 1000 milliseconds:

```
; Example3.scr
; Script file to display a series of slides.
expert 5
filedia 0
vslide Single Door
delay 1000
vslide Double Door
delay 1000
filedia 1
rscript
```

11.8 CUSTOMIZING LINETYPES

AutoCAD contains a multitude of linetypes such as continuous, hidden, centre, etc. including variants of the main ones available and these can be seen by typing linetype at the command prompt and then answering the next question with a ?. This will display all the line types that are integral to the AutoCAD package.

However, AutoCAD users may have their own preferred linetypes that may not be available from the database provided by Autodesk. In this case, users may wish to create ones that are peculiar to their own practice. This can be done by programming files with extension **.lin** in the text editor. To demonstrate this, an example line with a series of dashes and dots will be customized and created. Again, this can be done using a suitable text editor to code an ASCII file **egline.lin** as follows:

```
;;egline.lin
;;Linetype with a series of dashes and dots.
*EXALINE
A,15,-5,0,-5,5,-5,0,-5
```

The first two lines are simply comment lines to explain what the coding is for. The word EXALINE denotes the descriptive text of the linetype. The pattern of the line is started with the letter A, 15 denotes the length of the first dash. The line will be drawn similar to the way the pen plotter works. A positive number indicates that the pen will be down and drawing a line, the negative number indicates pen-up and therefore results in a blank space. The 0 indicates a dot or a point.

The description is only for the first portion of the line and the rest of the line is simply repeated with the same pattern. This can also be saved as an ASCII file in a suitable directory and the AutoCAD command prompt LINETYPE is entered followed by typing the name of the line, **egline.lin**. This will now be part of the database of lines available for use in the current drawing session. A sample line can be drawn to see the effect but if the dash-dot pattern does not appear, then it will probably be the linetype scale that needs to be adjusted to a suitable value.

11.9 CUSTOMIZING HATCH PATTERNS

In a similar manner to customizing linetypes, hatch patterns can also be customized to a user's preference in case the desired pattern may not be available in the AutoCAD database of hatch patterns. The same procedure as for linetypes can be used to customize or create a user's preferred hatch pattern. Using a suitable text editor, the following hatch ASCII file, **eghatch.pat** can be coded.

```
;;eghatch.pat
*EGHATCH, Example of My Own Hatch Pattern
45,0,0,0,15,5,15,0,-5
135,0,0,0,15,5,15,0,-5
```

The name eghatch.pat must correspond to that in the second line EGHATCH. The general structure of the hatch pattern is:

angle, x-origin, y-origin, delta-x, delta-y [dash-1, dash2...]

angle:	Angle at which the pattern is drawn to the horizontal.
x-origin:	x-coordinate from where the line is drawn from.

y-origin:	y-coordinate from where the line is drawn from.
delta-x:	Vertical offset between dashes in the direction that a pattern is drawn.
delta-y:	Separation between lines of dashes perpendicular to which a pattern is drawn.
dash-1/dash-2:	Pen motion similar to that for line types.

After saving the ASCII file, eghatch.pat, the text editor can be closed and then a rectangle can be drawn in AutoCAD, followed by the BHATCH command to select the internal area of the rectangle. The eghatch.pat can be accessed by using the User Defined or Custom hatch pattern on the Hatch dialogue box and then applying it on the rectangle. The other settings are as defined previously, where the scale and hatch angle can be adjusted to a suitable or desirable display.

11.10 CASE STUDY FOR A POWER AND ELECTRICITY DISTRIBUTION COMPANY

The data, modelling and analysis in this section are a summarized version of a research/case study carried out at an electricity distribution company which was computerizing and customizing its CAD system (Nyemba & Mbohwa, 2016). The research was presented at the *2016 International Conference on Industrial Engineering and Engineering Management* in Bali, Indonesia and subsequently published as a paper in the IEEM/IEEE Proceedings in 2016, from where more details can be obtained.

CAD systems are normally bought off-the-shelf and used in wide disciplines of engineering and other fields. However, due to the diversity of such disciplines, developers have to take into consideration, the requirements of different users for the efficient production of models and drawings. As such, most of these packages are provided in a generalized form but supplied with facilities to customize them to user requirements. Research carried out at a power and electricity distribution company in Zimbabwe revealed delays in attending to power faults and installations at their substations partly because of the non-availability of required drawings on time to attend to the faults. At the time, the company was also migrating from manual drawings to CAD, resulting in some cases of duplications and delays.

The company sought the assistance of CAD experts to develop an in-house strategy to customize their AutoCAD software through an industrial engineering approach to reduce man-hours and increase throughput by the generation and storage of their commonly used symbols as blocks that can easily be retrieved and used in drawings. This resulted in the reduction of lead times and attending to faults timely through the integration of their design and maintenance management systems. Purchasing CAD systems off-the-shelf tends to be cheaper than developing one in-house due to costs (Vinodh et al., 2010). With the rapid changes in technology, whichever option is chosen, continuous development is a necessity, hence why some organizations have preferred to leave this to dedicated software developers.

Although purchasing off-the-shelf packages can be equally expensive, coupled with annual licenses and upgrade costs, in-house development is also prohibitive especially if the company does not have skilled software developers. This has pushed many companies to opt for Commercial Off-The-Shelf (COTS) software or turnkey systems by purchasing complete systems from vendors (Ahola et al., 2008). In any case, most such systems are now supplied with facilities to customize them to a company's requirements. Apart from being provided with warranties, any special requirements by any organization can be requested from the dedicated software developers, and usually these requests are incorporated in the ever-changing versions of the CAD software systems.

In order to remain in business and to ensure commercial viability, software developers such as Autodesk, supply their systems in a generic form suitable for a wide variety of customers. Most of these developers have gone further to develop variants of their main CAD software to suit different applications such as Architecture, Pipe Design and Instrumentation. Benefits for developing in-house software are usually realized in long term and these are not viable or justifiable for the sizes of businesses in industrializing countries such as where this case study was carried out. Hence, most companies end up balancing the two, purchasing a generic system and in turn customizing it to their requirements (Hsu et al., 2014), which is exactly what the power and electricity distribution company did.

The need to customize CAD software arises from many reasons such as inadequacies in the generic software or very specific and narrow design portfolios by companies that may not be sufficient to share with other companies, as well as the capital-intensive nature of custom or discipline-specific built software such as AutoCAD Architecture. However, it must also be borne in mind that custom-developed or discipline-specific software is more flexible and enables the growth of the system in line with the business resulting in efficiency, better throughput and quality of service (Basoglu et al., 2009).

Due to their nature of operations in distribution of electricity and maintenance of their substations, the case study company makes use of and produces a wide variety of engineering and technical drawings. Their Protection Drawing Office (PDO) was responsible for the development of all drawings and models. In line with rapid changes and global competition, the company had to keep abreast with world trends, thus their decision to computerize and customize the PDO for consistent and accurate drawings as well as timely attention to faults, driven by the demand from their various substations. As such, the company was not only faced with transition and computerizing the whole organization but also the need to customize their newly acquired CAD software, apart from skilling their staff to operate the CAD system.

As alluded to in previous chapters, most CAD users do not realize the full benefits of their systems because they mainly just concentrate on the drafting capabilities but very little on modelling and other facilities such as blocks and 3D visualization (Quintana et al., 2012). The case study company employs over 10,000 workers in its mainly four engineering subsidiaries, thus making it a large enterprise by the country's standards. As various sectors of the economy in the country have been growing, so has the increased demand for electricity, implying more substations throughout the country. This also implied the increased need for technical support and hence the need for the rapid production of engineering drawings and models by its subsidiaries. This was the basis on which this research and case study sought to balance COTS software and developing an in-house solution or strategy, leading to the adopted approach of Customized Add-on In-House Development (CAID) based on both strategies (Nyemba & Mbohwa, 2016). The balance needed to be carefully managed and this was achieved through treating the development of the CAID software as an integral part of the company's development strategies as a whole. This case study therefore focused on the development of CAID superimposed on their AutoCAD software.

11.10.1 Customization strategies adopted

The industrial revolutions and transformations in industry have resulted in rapid changes in technology, equally so in rapid changes in software development evidenced by steep rises in changes by Autodesk since their first version in 1982 (Kushwaha et al., 2012). Software developers have adopted several strategies to upgrade their software through avenues such as agile software development (ASD) in which the software is adjusted through iterations, using methods such as Dynamic Systems Development Model (DSDM) or Scrum and Extreme Programming (XP) (Misra and Singh, 2015). ASD was used to resolve problems commonly encountered in traditional software development such as the limitations of and difficulty in changing the users' existing software to incorporate the additional changes. However, this approach has active user involvement and close collaboration during the development cycle. This invariably implies that users have to be devoted full time to this and this can be quite costly for most organizations, although by nature, ASD is agile during the development process and thus flexible to transition and change direction.

As demonstrated in the earlier part of this chapter, AutoCAD was developed as an open architecture software and Autodesk continues to improve this in such a way that the user does not have to leave the AutoCAD environment in order to make customization changes but all can be done within the same environment.

Users can also add their own menus to the AutoCAD pull-down menus which can be used to load list processing (LISP) routines and so forth [15]. In addition, menus can be used to launch favourite or frequently used commands and to execute macros. Equally, AutoCAD allows users to develop their own macro-programs through AutoLISP and script files to automate operations. ActiveX Automation in AutoCAD also provides ways to manipulate AutoCAD through programming from within or outside AutoCAD by enabling various AutoCAD objects to the 'outside world'.

Enabling these objects allows many different forms of programming languages or platforms such as Word, Excel etc. to access them. The use of the Design Web Format enables others to view drawings and models without the need to purchase AutoCAD but through the Internet. Such a platform is an area that has increased usage in industrial engineering and the marketing of engineering products (Wang et al., 2013). AutoCAD also allows users to improve the interface between the user and the system through the Dialogue Control Language (DCL), a platform used to create dialogue boxes for easing communication and inputting of information. The case study focused on utilizing all aspects covered in this chapter in order to address the company's failure to timeously attend to breakdowns at their various substations and projects throughout the country, purportedly due to delays in the provision of the requisite drawings. Being a relatively large company, and the sole distributor of electricity to all industries and residential areas, it was prudent to ensure that CAID was developed with minimal interruptions to operations. Hence, the use of ASD was shelved in preference for customizing the AutoCAD that the company was already in possession of.

The major functions of the company revolved around the provision of electricity driven by their major departments in mechanical (generation) and electrical (distribution) engineering, hence the need to ensure that equipment was well maintained to reduce occurrences of faults. The PDO at the company is responsible for designing security features that guard against system overloads, hence the need to be well resourced and maintained. Drawings produced in the PDO are mostly schematics showing design and installation details for the wiring at substations. There is a multitude of such electrical drawings throughout their substations, albeit most of them are similar, consisting of standard symbols. Bolstered by their new computer system that records and transmits faults to the national control station, CAID could also benefit from the setup, so that technical drawings and changes required to attend to faults can be managed through their Maintenance Management System (MMS) Physical Assets Module.

The development of CAID needed to take into cognizance, the need to automate the creation of symbols in order to produce drawings rapidly, relative to their manual and inefficient use such as the insertion of blocks. Apart from delays in producing the required drawing, this also demanded fast processing computers. Non-graphic information such as specifications and bills of materials associated with the electronic components could not be retrieved automatically, hence the need to incorporate this facility in the CAID. Any new equipment acquired by the company was commissioned by the PDO and documentation for such was filed accordingly as 'As-Built'

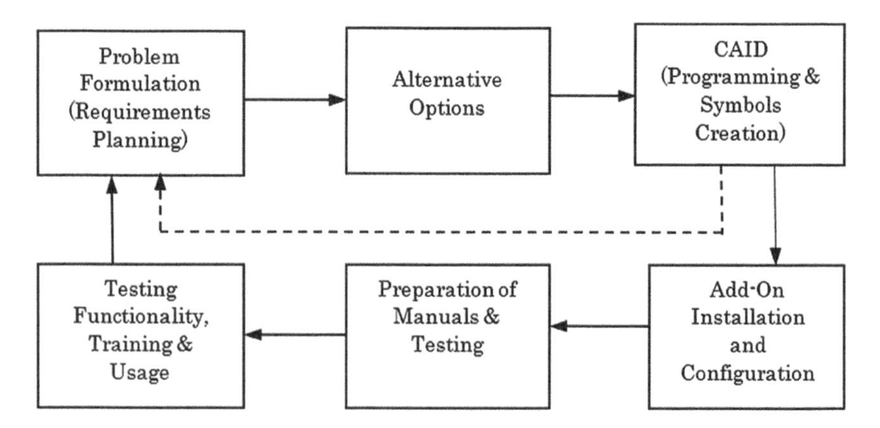

Figure 11.4 CAID development cycle strategy.

drawings. It therefore made sense that the PDO would be the first to the MMS Physical Assets Module to enable the compilation of a database for assets, inclusive of an on-help facility.

The planning for the development of the CAID was derived from working directly with the engineers in the PDO to tap from their experiences and how they traditionally carried out their tasks, albeit manually. Figure 11.4 shows a summarized cycle of how the CAID was developed at the company. Checks and balances as well as continuous test runs were carried out throughout the research period as customization of the software was in progress. Continuous improvements, shown by dashed lines in Figure 11.4, were also made in relation to how they conducted their business manually.

11.10.2 Results obtained

Several alternatives that were required to develop the add-on facility were considered such as the use of toolbars, pull-down menus, dialogue boxes as well as blocks and attributes. Eventually, a combination of all these was adopted, with the two main programs to drive most of the routines in the CAID and to drive the dialogue boxes written in AutoLISP. An online help facility was also incorporated in conjunction with each dialogue box that was created. Alongside the development of dialogue boxes and customized routines in the CAID, the commonly used electrical symbols by the company were also created as blocks and stored in selected databases such as Civil, Electrical, Mechanical etc. for ease of retrieval, depending on the application and whether it was for generation or distribution.

The alternative solutions were derived from a morphology chart which consisted of procedures for the automatic insertion of blocks and for attaching attributes and specifications to the blocks in order to identify the components represented or to store maintenance data associated with particular equipment. Secondly, the CAID system also allowed maintenance information to be attached while the facility customized the extraction of all non-graphic information such

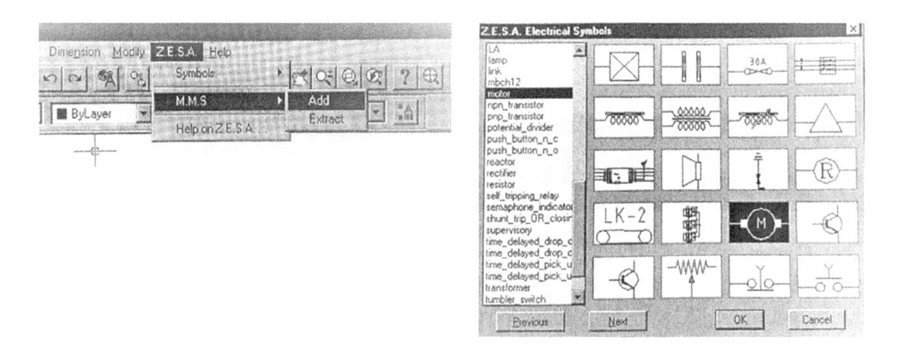

Figure 11.5 CAID add-on utility and electrical symbols created.

as maintenance data in a drawing for use in the MMS Physical Assets Module. Figure 11.5 (left) shows a snapshot of the pull-down added under the CAID as well as some of the symbols that were created and stored as image slides (right).

The pull-down menu created as shown in Figure 11.5 (left) facilitated the insertion of the company's commonly used symbols also shown in Figure 11.5 (right) from the respective databases into any drawing, using the image tile menu as shown. The DCL was used to create an additional dialogue box as shown in Figure 11.6 (left) to prompt the user to enter attribute values for each electronic symbol such as a transformer. If the part selected for insertion required maintenance data, then the dialogue box shown in Figure 11.6 (right) allowed the insertion of details into the Physical Asset Register. The precise location and size of the block on the drawing are controlled by the Insertion/Base Point (x, y, z) coordinates, Rotation and Scale on insertion, complete with all the non-graphic information such as specifications. The attributes, the definition and insertion of such information for each symbol on inserting into a drawing are the same as what was covered in Chapter 7. The same procedures were used to automatically define selected entities as blocks, and then attach maintenance attributes to the block, before re-inserting the block in the desired position. The **Extract** option under the MMS submenu (Figure 11.5 left) launches the customized

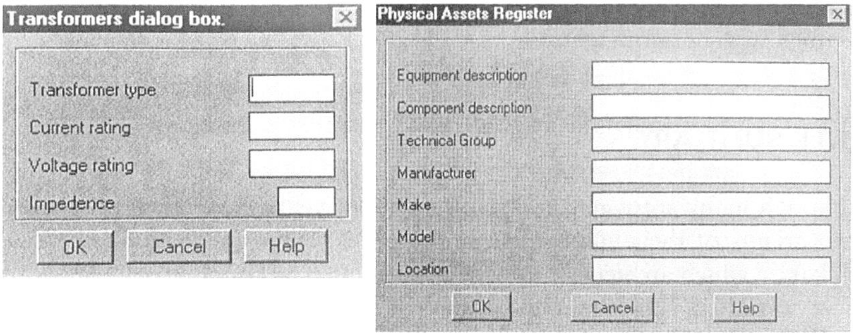

Figure 11.6 Attributes dialogue box and physical assets register.

attribute extraction command. Also, as described in Chapter 7, the attributes are exported to a space delimited text file that can then be utilized as input to the MMS Physical Assets Module, which is based on a suitable database such as Microsoft Access.

11.10.3 CAID accomplishments

The development of the CAID at the power and electricity distribution company was largely successful as it enabled the integration of AutoCAD with the company's MMS Physical Assets Module in the form of provision of non-graphic information such as bills of materials. In addition, the system also allowed the company to standardize their drawings for use at their substations throughout the country. There were also some marked reductions in reworks particularly where maintenance data had traditionally been entered manually, thereby causing delays in attending to faults. Engineers and technicians found the CAID system user-friendly as they no longer needed to refer to handbooks and manuals, which was time-consuming and prone to errors.

The advantage of storing and retrieving blocks with their attributes was almost like halving the time required relative to what the engineers and technicians did manually. The CAID also standardized the definitions of attributes in drawings thereby only requiring the values of attributes to retrieve blocks from databases, and the automatic compilation of non-graphic information to spreadsheets of bills of materials for the purposes of forward planning and scheduling maintenance. However, it was recommended that their engineers and technicians needed to be continually trained as often as new versions of the AutoCAD software were upgraded, thus also requiring upgrades in the CAID add-on utility.

The major limitation or challenge for this sort of work is its sustainability in view of rapid changes in technology, particularly the rate at which Autodesk has been upgrading its software. This ultimately also requires adjustments in the CAID utility, as routines change and so should the CAID. A dedicated team specializing in upgrading the add-on utility may be necessary to cope with upgrades from Autodesk and this could be costly. Further work into improving the CAID may be necessary to establish a sustainable system that can cope and work in tandem with changes in software upgrades, coupled with continuous training.

11.11 SUMMARY

Although many software developers in recent years have taken to developing variants of their main software to suit different applications, the main packages, which are generic in nature (catering to all disciplines), remain the most sort-after. This is mainly because they are more affordable as they are in demand from many users. Generic CAD software such as AutoCAD is provided with facilities to enable users to tailor-make it to their needs through customization, apart from the ability to program script files to automate tasks, in order to improve the efficiency in the production of drawings and models.

This chapter focused on guidelines to assist AutoCAD users to customize their software to specific preferences, ranging from removing, adding or adjusting menus and toolbars. AutoCAD users have the option to reduce or completely remove some of the pull-down menus from the menu bar and at the same time to create new toolbars and menus which they can associate with selected commands. Icon image boxes can also be created, in which slides of symbols that are commonly used can be added for easy retrieval and use in drawings.

This chapter not only covered how customization was carried out in the earlier versions of AutoCAD up to Release 12 but also how Autodesk has simplified the processes in the newer versions of AutoCAD where most of it is now done within the package through dialogue boxes. To demonstrate the usefulness and importance of customization to organizations, a case study for a power and electricity distribution company was presented. In this case study, an add-on utility, CAID was developed to assist and enable the company to customize their CAD software as well as to automate tasks in the usage of blocks and attributes as well as generate non-graphic information such as spreadsheets of bills of materials. This greatly improved the company's operation by reducing lead time to attend to faults as well as productivity and efficiency in the development of drawings and models for their substations.

11.12 REVIEW EXERCISES

1. Explain the general procedure for customizing menu bars and toolbars, paying attention to how command icons can be added or removed from a particular menu or toolbar.
2. Most CAD software developers provide many facilities to tailor-make their packages to suit user requirements. List and explain at least five of such facilities available in AutoCAD, providing details of what benefits organizations can enjoy from such customizations.
3. In recent years, CAD software developers have tended to develop variants of their generic software to cater for specific applications. List five such variants that Autodesk has developed over the last few years and explain how these variants are different from the generic software.
4. Outline the major benefits of customizing CAD software for companies and explain how Autodesk has simplified customization through the migration from the earlier versions of AutoCAD to the later ones.
5. The use of script files can enhance the application and use of CAD packages. Outline at least five such uses where script files can be used to automate tasks in AutoCAD.

Chapter 12

Management of models and drawings for output

12.1 INTRODUCTION

The penultimate aim for developing skills in operating CAD systems is to be able to produce drawings and models in an efficient and presentable manner. The skills covered since Chapter 1 include the ability to draft and model in 2D and 3D, the use of attributes and blocks as well as how to customize AutoCAD in order to suit particular operations or specializations. While some guidelines have also been provided to manage the creation and use of drawings and models developed from CAD, this was done for certain and specific applications. The skills acquired may not be very useful unless users are able to adequately manage and present their drawings and models in an attractive and business-oriented manner.

This chapter focuses on guiding users on how to package their drawings and models to management or potential customers for decision-making, especially where new products are being introduced. The packaging should be done in such an attractive manner as to persuade management to give the project a go-ahead or for customers to place orders. The usual practice in the past was to draw manually on paper and then store the papers in some cabinets for later use on construction sites, production plants etc. or in the plant. This was costly in terms of paper, storage and handling.

Nowadays, with CAD, printouts on paper are no longer really necessary since models and drawings can be shared on laptops, mobile phones and other devices. This chapter will also guide users on how to create suitable drawings and models for use in different such places as construction sites, production plants, substations etc. as well as organizing the drawings using viewports, model space and paper space. For those still making use of printouts on paper, guidelines will also be provided to organize the drawings and models for output on different sizes of paper.

DOI: 10.1201/9781003288626-12

12.2 MODEL AND PAPER SPACES IN AUTOCAD

Generally, models and drawings developed in AutoCAD are created in model space, the default environment that is loaded when AutoCAD is started. On the other hand, paper space is the AutoCAD environment in which drawings are organized or laid it out in preparation for printing, usually on paper. Such models and drawings also need to be organized for display in different formats such as portable document format (PDF), for use in production plants and construction sites, in their electronic form. Although traditionally drawings were produced on paper and even during the earlier days of CAD, after modelling on computer, they were also produced on paper, nowadays, with all the multitude of formats available, it may not be very necessary to produce such models on paper. Most CAD systems now have facilities for easily converting models to different forms that can be used on different devices at construction sites and production plants.

CAD users are encouraged to use a full scale when creating drawings in model space. This was not possible with manual drawings because of the limitations of paper sizes available and in use. As such, models in CAD should be generated using full scale while scale adjustments can be made in paper space in preparation for printing. Paper space, commonly referred to as **Layouts** in AutoCAD, allows users to layout drawings ready for printing.

At the bottom left-hand corner of the AutoCAD screen, as shown in Figure 1.7, there are three options that are loaded when AutoCAD starts, the Model Space and second layouts, **Layout1** and **Layout2**. The default environment for setting out and creating a drawing will be under **Model**. Having completed a drawing or model, users can choose to lay out the different parts of the drawing using any of the Layout options. These are not limited to two but more layouts can be added by clicking the (+) sign after Layout2 depending on how many different styles the user wishes to organize their drawings, as will be seen in the guided examples to follow in this chapter.

While Paper Space is specifically meant to organize drawings for printing, any model or drawing under Model Space can still be printed directly from that environment. This is usually done for simple layouts containing only a few views and this is usually done to save the time required to organize models and drawings in paper space. However, Paper Space is generally the best way to lay out models in preparation for printing as it allows the setting up of several views of drawings in one paper using viewports. More specifically, more views of the same drawing in different scales can be laid out in separate viewports.

12.3 VIEWPORTS

Viewports are a facility in CAD systems such as AutoCAD where users can split the usual AutoCAD screen into several mini-screens to allow for the insertion of different views of a drawing or model under development. Such a facility allows users to create these viewports for different views, taken from different angles or different scales, in the Model Space to enable them to be displayed on Paper Space ready for printing. As such, viewports are display windows similar to the entire AutoCAD screen. These can be created using the View pull-down menu, followed by Viewports and then selecting the number of viewports required, as shown in Figure 12.1.

While several viewports can be created such as the four shown in Figure 12.1, only one viewport is active at any given time. After creating the viewports, the active viewport will be the one where the AutoCAD cross-hair will be active, while the other viewports will be inactive and will instead have the normal Windows cursor. The active viewport is the one where all changes and additions are possible. To activate any viewport, the LMB can be clicked within the desired window. It should also be noted that viewports can be created both in Model Space and Paper Space. While several viewports can be created within the main AutoCAD screen, even smaller or additional viewports can also be created within desired viewports, using the same procedure as outlined above. Whereas only one viewport can be active at any given time in Model Space and can be controlled separately, the same viewports in Paper Space can be controlled at the same time so as to easily manage them to enable the organization for printing purposes. Viewports are particularly useful to display different elevations or 3D models on one screen for clarity and thus reducing complexity.

Elevations or models in viewports can be magnified (zoomed) up or down using the mouse roller, just like the normal modifications carried out and explained in Chapter 3. If views in viewports are connected, such as elevations of an object (plan, front and side etc.), any changes effected in one viewport will automatically update the connected views in the other viewports, even if only one viewport will be active at any given time. Such changes include the scaling of objects in the active viewport. However, it should be noted and encouraged that scaling for the purposes of preparing a drawing or model for printing should only be done in Paper Space, to reduce complexities in model representations in CAD. The visibility of layers in viewports can also be controlled in the same manner as discussed in Chapter 5 and this particular aspect is important in the organization of drawings for printing, as will be demonstrated in the following sections.

12.4 MODELLING CAST IRON BASE

The modelling of a cast iron base will be used as a worked-out example to demonstrate how model space, viewports and paper space are used in AutoCAD to prepare and organize a model for printing. In addition, two components will be added in the form of an assembly. Thus, it is not just an example to demonstrate the organization of models but also the use of

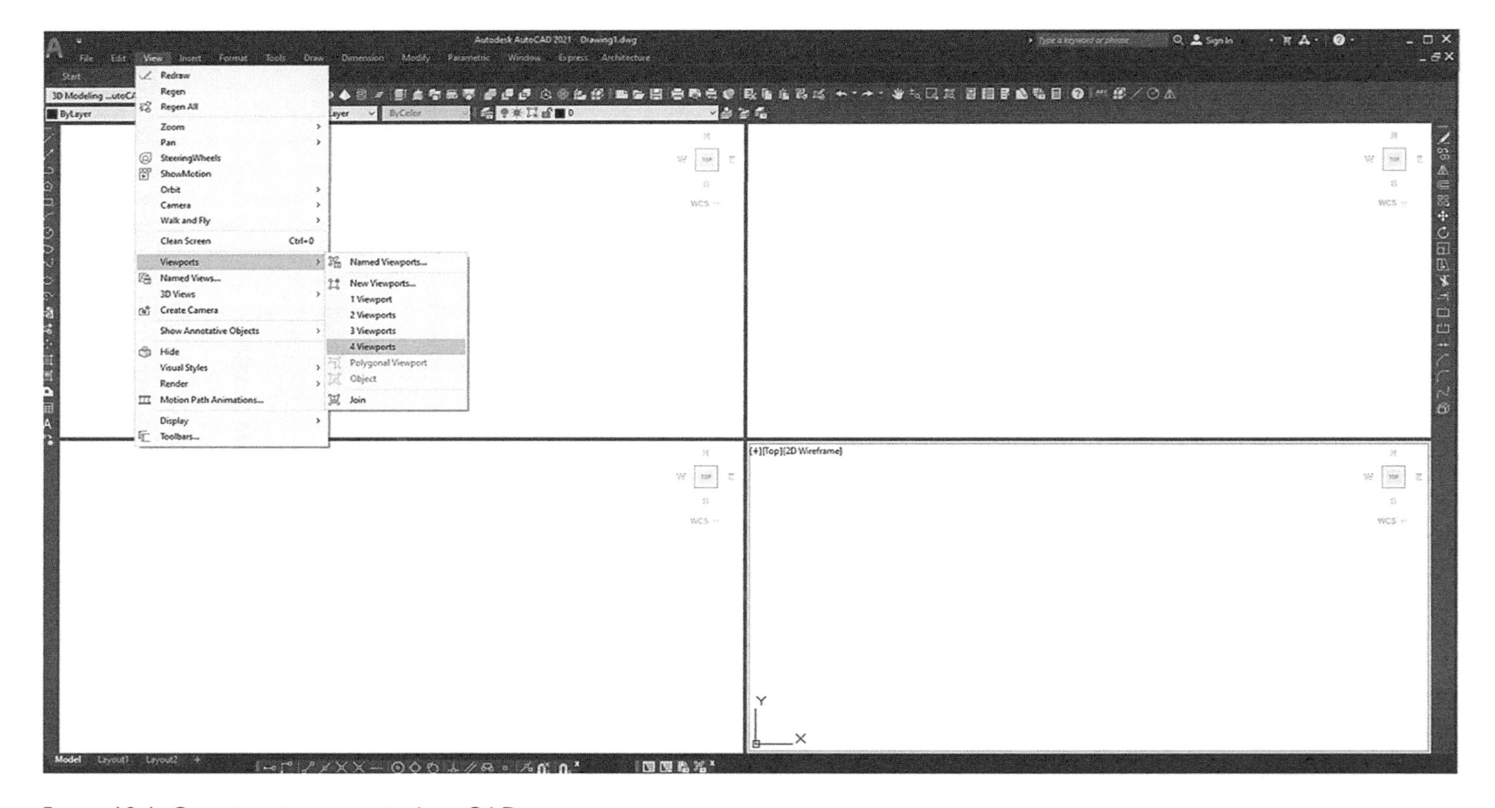

Figure 12.1 Creating viewports in AutoCAD.

constructive solid geometry (CSG) and assembly as covered in previous chapters. The three orthographic views of the cast iron base are shown in Figure 12.2 (left). The 3D model of the cast iron base which is required to be generated from the orthographic views is also shown in Figure 12.2 (right). The dimensions provided alongside the three orthographic elevations, Front, Plan and Right Side, in third angle projections, are sufficient to completely describe the cast iron base. It should be noted that at this stage, the object is a single cast component consisting of a base with chamfers on two edges as shown, as well as fillets on the edges joining the base and middle section, with cylindrical protrusion with a through hole.

The first step in CSG modelling would be to study the component carefully and then decide on three issues: most optimum viewing angle or orientation (viewpoint), the optimal number of solid primitives required and how these primitives are combined to get the resulting model. Seeing that the Front, Plan and Right-Side Elevations were provided, a suitable viewing point could be (1, −1, 1) or more commonly the SE Isometric for later versions of AutoCAD. This orientation is the most suitable in that the user will be able to see the three faces, Front, Plan and Right Side.

The optimal number of solid primitives required to create this model would be two boxes, one on top of the other and two cylinders, the outer protrusion and the inner one which will be subtracted to create the internal hole. The chamfers and fillets on the edges can be created by subtracting two additional primitives such as cylinders and wedges but the most optimal route would be to create the solid chamfers and fillets on those edges using available AutoCAD commands, chamfer and fillet. As such, only two primitives are required, box (2 off) and cylinder (2 off). Assuming these are labelled B1, B2, C1 and C2 as shown, then the Boolean operations required would be (B1 + B2 + C1) − C2.

Using the dimensions provided on the orthographic elevations, the first box (base) can be created using the BOX command, length 200, width 100 and height 40. This can be done by setting the first corner to (0,0), the other corner will thus be (200,100) and the height is 40. The BOX command will then be repeated to create the smaller box on top of the larger one, ensuring that the coordinates are correct, according to the dimensions on the elevations. In this case, the first corner will be (30,0,40), the other corner (170,100,40) and a height of 20. When constructing such primitives using absolute coordinates, it is important to ensure that dynamic settings (DSETTINGS) have ABSOLUTE COORDINATES checked on, otherwise this may give you some unexpected errors. This is shown in Figure 12.3 (left).

The next step would be to create the two cylinders, firstly C1 of height 60 and radius 40, located centrally on top of the smaller box. To ensure that the cylinder is sitting centrally, a construction line can be drawn from the midpoint of the smaller to the midpoint of the other end, then using the midpoint of this construction line as the centre of the larger cylinder. The CYLINDER command is then used to create the larger cylinder, using the three parameters

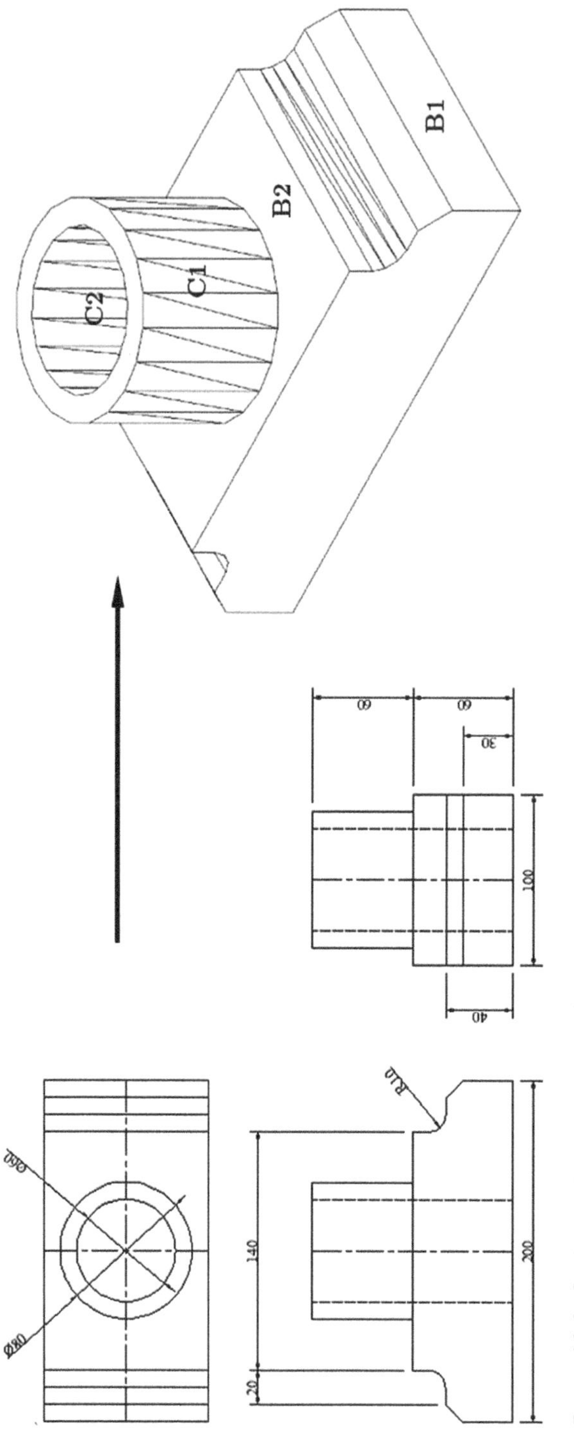

Figure 12.2 Orthographic elevations of a cast iron base and its 3D model.

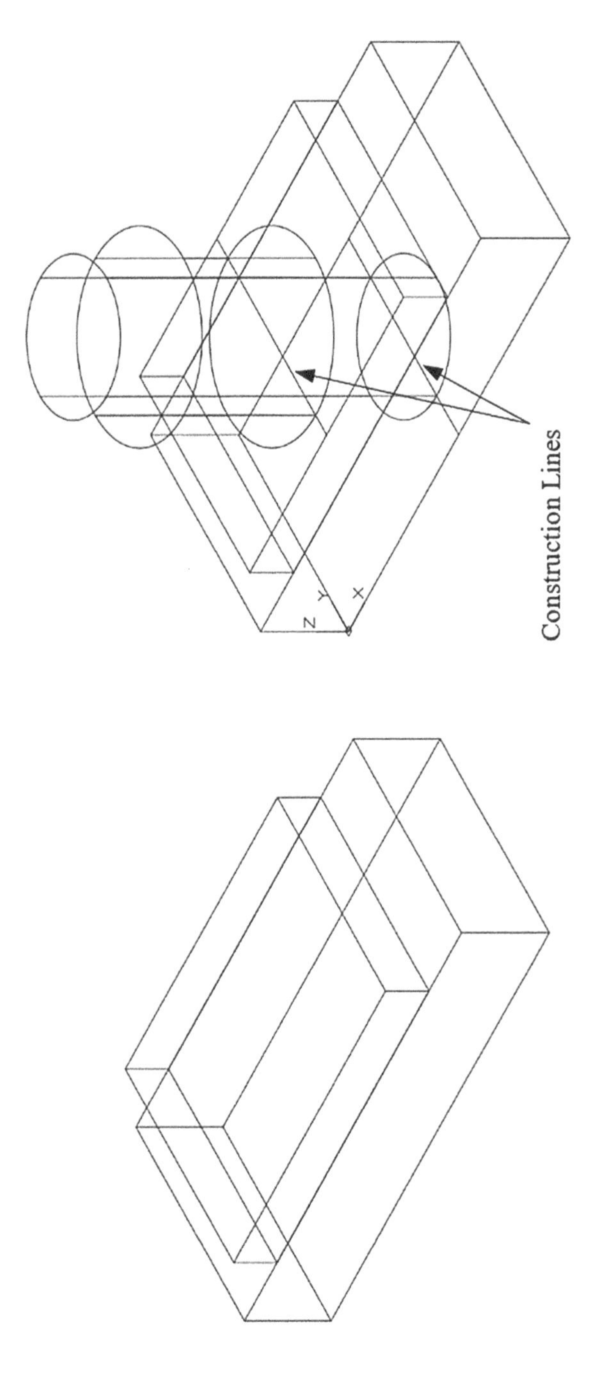

Figure 12.3 Solid primitives for the cast iron base.

above. Secondly, the smaller cylinder, C2 of radius 30 and a height of at least 120 (sum of the heights of the two boxes and large cylinder since we need to subtract it in order to create a through hole). Even if the height is 140 or 150 it does not matter as the objective is to subtract this, remaining with a hole of the height of 120. The small cylinder is located centrally on the base of the larger box and can be created by first creating a construction line as in the previous case and then using the midpoint of that line as the centre of the small cylinder. Using the three parameters of radius, height and centre of the cylinder, the small cylinder is also created using the CYLINDER command. Although absolute coordinates can also be used to locate the cylinders, the midpoints of the construction lines may be easier to visualize as the construction progresses. The construction lines can be erased afterwards to reduce complexity. The resulting display with the cylinders is shown in Figure 12.3 (right).

Using the annotation in Figure 12.2, a single component can be created by uniting the two boxes and the larger cylinder, then subtracting the smaller cylinder from that union, that is,

$$B1 + B2 + C1 = D \tag{12.1}$$

$$D - C2 = E. \tag{12.2}$$

The resulting display € is shown in Figure 12.4 (left) with hidden lines removed by using the AutoCAD command HIDE. The original mode with all wireframes including hidden lines can be restored using the REGEN command to regenerate the model. Chamfers and fillets in solid modelling are dealt with in much the same way as for 2D drafting and modelling. From the orthographic elevations, two chamfers need to be created on two edges of the larger box and two fillets also need to be created on the edges adjoining the two boxes, on both sides.

Starting with the chamfer, using the CHAMFER command, the edge where the chamfer is required is selected then and entered (OK) followed by setting the chamfer distances of D1 (10) and D2 (10) as provided on the orthographic elevations. The same edge is selected when prompted for the edge to chamfer and the chamfer will be created. This can be repeated on the other side but in real practice, the two edges can be selected and chamfered at the same time.

The FILLET command is used for the two fillets on the edges adjoining the two boxes, which are the same edges to be filleted with a fillet radius of 10 as provided on the detailed orthographic elevations. The command will prompt again almost in the same manner as the chamfer, thus selecting the same edge. Similarly, the two edges can be selected and filleted at the same time to save time. The resulting display is as shown in Figure 12.4 (right), which is the same as the 3D model in Figure 12.2 at the beginning of this worked-out example.

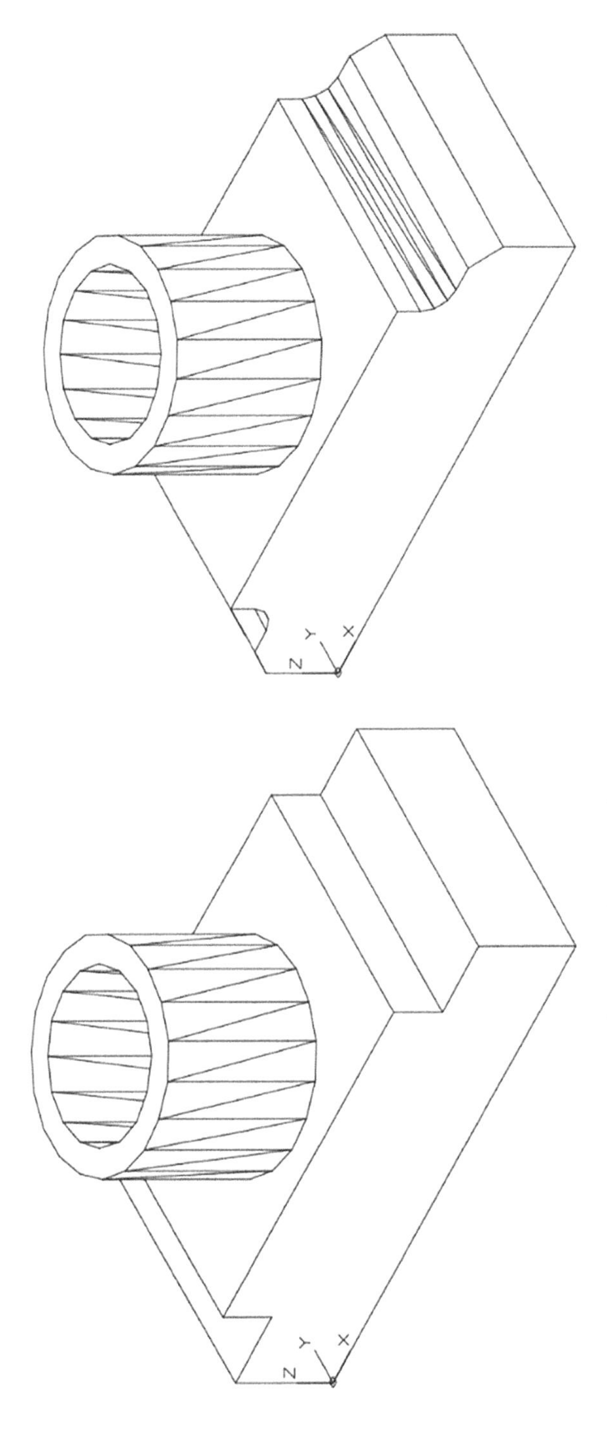

Figure 12.4 Cast iron base model (before and after chamfer and fillet).

12.4.1 Model space viewports

Figure 12.4 shows the complete model of the cast iron base as a single component combined using CSG. However, there are two additional components that will be assembled and attached to this base, that is, a shaft going through the hole and an O-ring outside the larger cylinder. The next stage in organizing the model for output on paper would be to define some layers over which the cast iron base, additional components, title block and viewports will be placed. Five layers that can be distinguished by colours such as **Cast Base** (Red), **Shaft** (Yellow), **O-ring** (Default-W/B), **Viewports** (Default-W/B) and **Title Block** (Default-W/B), can be created. The cast iron base that has been created already can be changed and assigned to its correct layer and immediately, the changes in properties will be visible in the cast iron base on the screen.

Before proceeding to assemble create and assemble the two components, the current layer should now be changed to **Viewports**, followed by splitting the screen into four viewports using the **View** pull-down menu, viewports and selecting four viewports to result in a display such as the one shown in Figure 12.1. The four viewports are assigned different elevations as follows:

Top left viewport	—	Plan view
Bottom left viewport	—	Front elevation
Bottom right viewport	—	Right side view
Top right viewport	—	Conceptual or realistic 3D view

The four viewport displays are changed one at a time, firstly by activating the top left viewport and then using View pull-down menu, followed by 3D views and Plan. The top left viewport will thus be automatically converted to the Plan. The procedure can be repeated for the next two viewports until all the required orthographic elevations are in place. The last viewport (top-right) will remain as the wireframe model of the cast iron base. When the viewport displays are converted to orthographic views, the resulting sizes in terms of magnification may be different from one viewport to another. At this stage, it is important to adjust and align these views accordingly by either zooming in or out on each viewport until all three viewports are correctly linked, as was discussed in Chapter 6. Further refinements such as changing hidden detailed lines etc. can also be made at this stage.

The next step would be to assemble or add the shaft and O-ring to the cast iron base. This can only be done with certainty by using the wireframe model in the top right viewport. Firstly, that particular viewport is activated and *Shaft* is made the current layer, followed by creating a cylinder with a radius of 30 and a height of 200 with its bottom centre coinciding with the bottom centre of the hole in the cast iron base. The O-ring is made of the current layer and a torus (radius 50, tube radius 10) on top of the flat surface of the base wrapping the outer cylinder is created. As these are added, details are also updated and added to the other three viewports and the 3D model can now be changed to a conceptual view as shown in Figure 12.5.

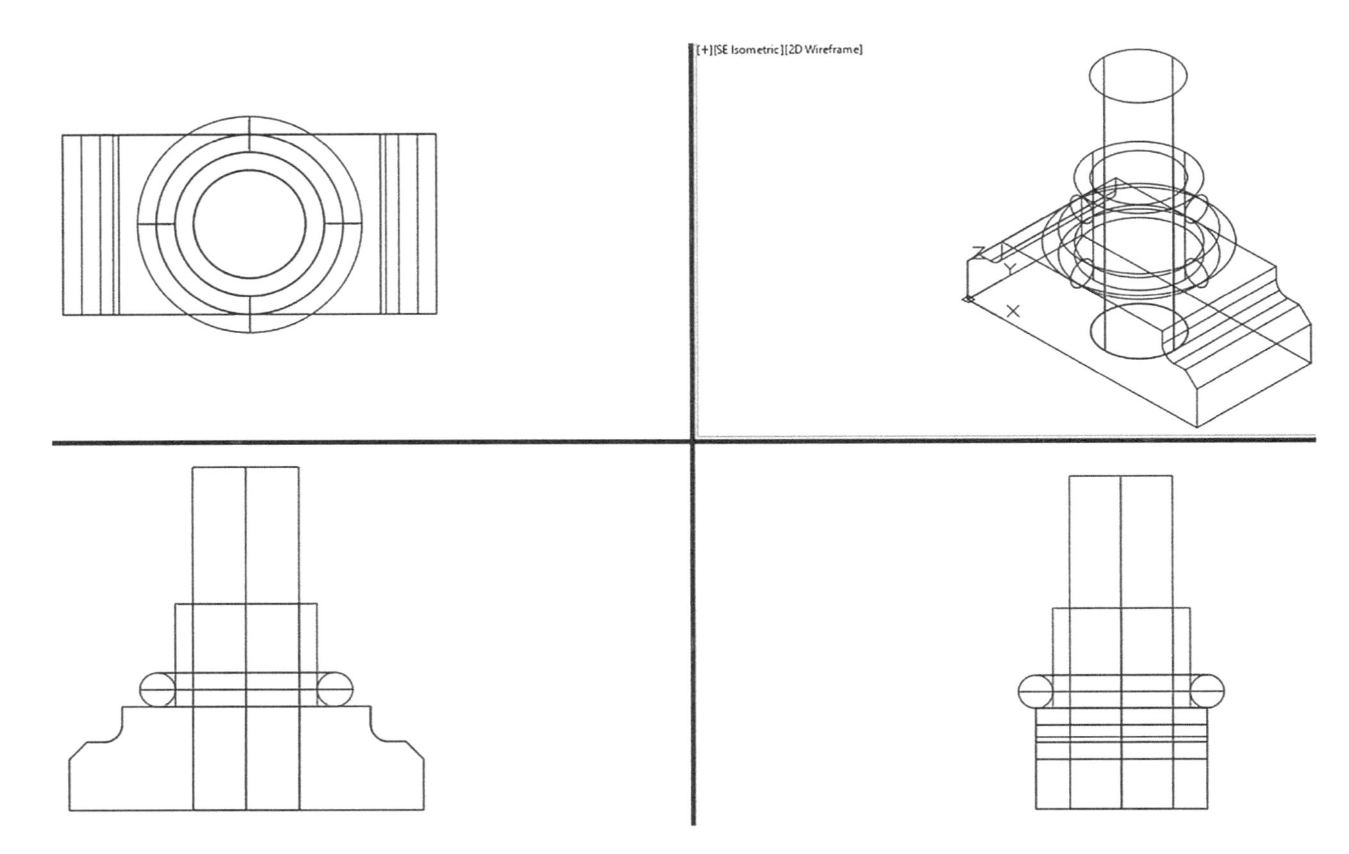

Figure 12.5 Model of the cast iron base assembly in four viewports.

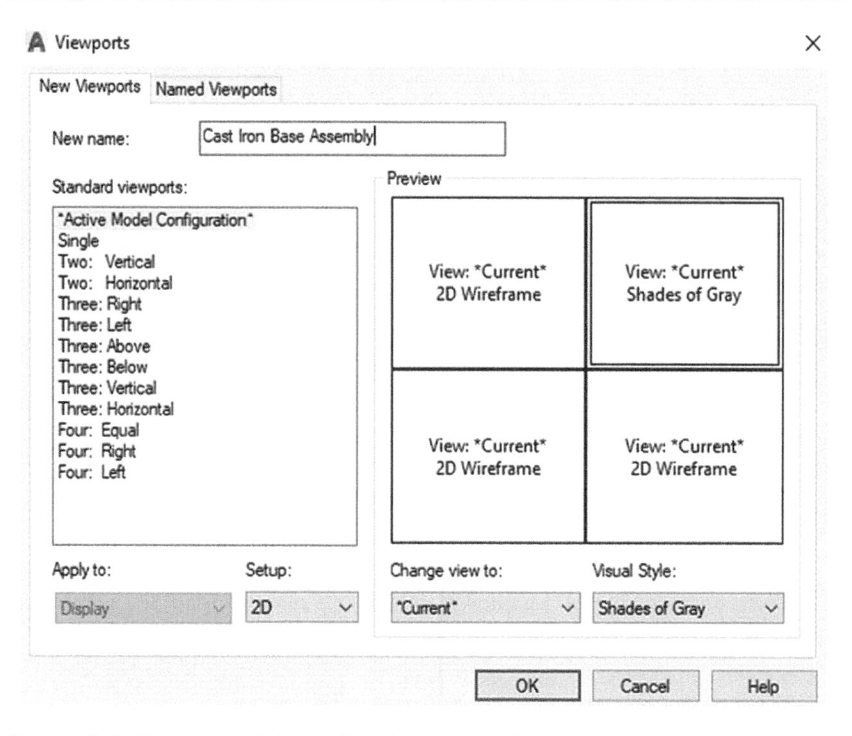

Figure 12.6 Viewports dialogue box for creating/saving viewports.

The model can now be saved by using the VPORTS that brings up a dialogue box (Figure 12.6), where the name Cast Iron Base Assembly can be entered and saved. Although the model space viewports are complete, they can only be printed or plotted, one viewport at a time as the output device only recognizes the active viewport, hence the need to go further to organize the models in paper space to enable them to be printed together as displayed.

12.4.2 Paper space viewports

To migrate from AutoCAD model space environment to the paper space environment, Layout1 at the bottom left of the AutoCAD screen can be used. Alternatively, the tilemode variable can be set to zero (0). A while background appears, appropriately representing the paper for printing. On the edges of that white background will be a dashed line, which basically represents the printable area. One of the viewports will also be displayed but at this stage, it does not really matter which one appears as this will shortly be erased as viewports are imported. Depending on what limits would have been set at the beginning of the model, AutoCAD simply uses the available paper size to suit those limits. However, before importing the viewports, the paper size for this model can be set to A3 (420 × 297) through the File pull-down menu, then Page Setup Manager to bring up the dialogue box shown in Figure 12.7 (left) and the second dialogue box (Page Setup – Model)

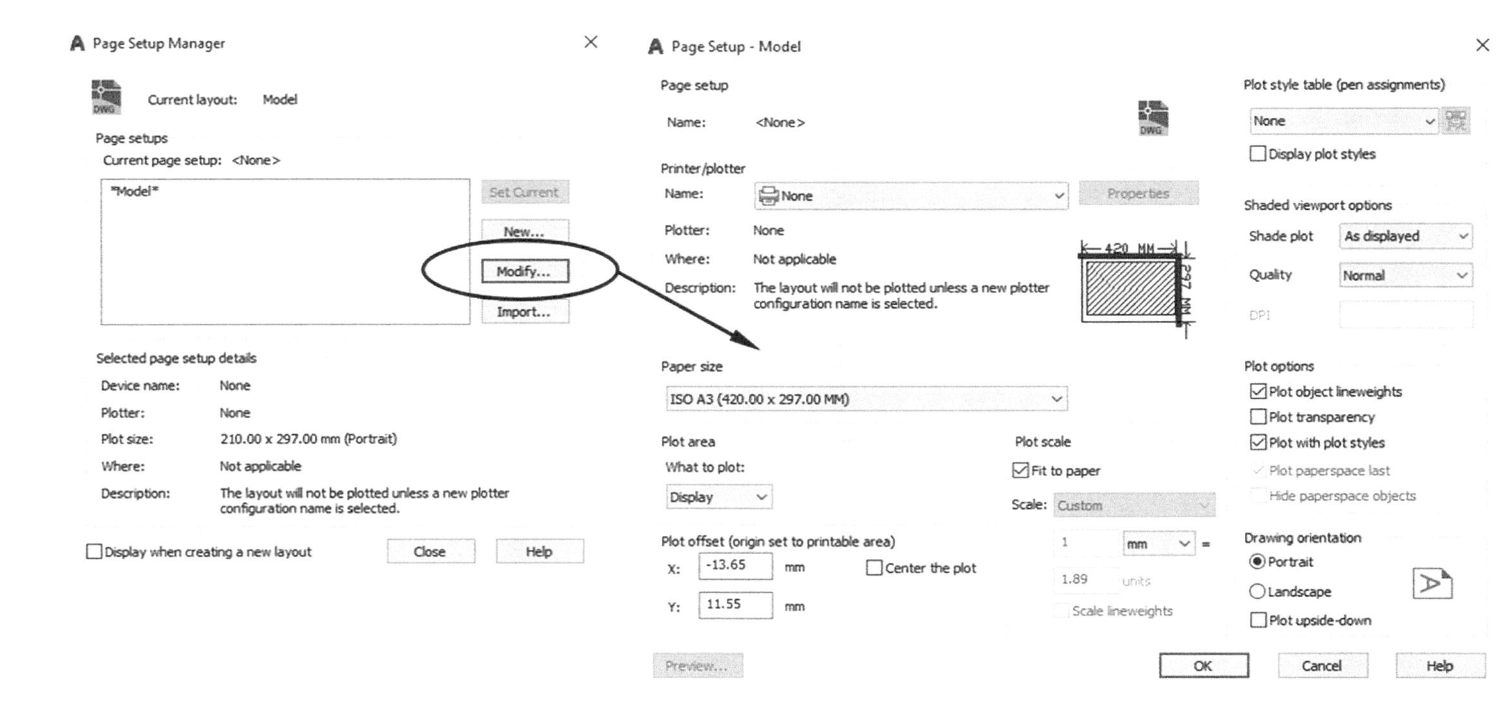

Figure 12.7 Page setup manager dialogue box.

obtained by selecting Modify on the first dialogue box. Several parameters can be set on the second dialogue box but for now, only the paper size needs to be changed to ISO A3 (420 × 297) as shown in the figure.

After setting the paper size to ISO A3, the paper space screen can be zoomed to ALL in order to see the entire paper and printable area (within dashed lines). The next stage would require that the user inserts their title block, which in this case, the one which was created in Chapter 2 and shown in Figure 2.7, will be used. The Title Block layer is activated to become the current layer. There are several ways in which the title block in Figure 2.7 can be imported into the current drawing in paper space, either opening up a new drawing with the title block, then copying and dragging into the paper space of the current drawing or using X-ref and inserting it into the paper space drawing as an attachment. However, the easiest and fastest way would be to open a new drawing with the title block as a template and copy all that is on the title block and drag it into the paper space of the current drawing. It might be necessary to do some adjustments to the title block to make sure that it falls within the printable area.

Before inserting the various viewports previously created, the Viewport layer is activated to become the current layer. At this point the viewport currently displayed can be removed by erasing it so that the entire paper space is free but only displaying the title block within the printable area. Using the View pull-down menu, followed by Viewports and Named Viewports, the previously saved Cast Iron Base Assembly will appear and then be selected for loading. The system will prompt for two corners within which the viewports can be inserted. In this case, these would be the lower left and upper right corners of the blank area of the Title Block and where necessary, the option, **fit to fill** the area with the four viewports is selected. Viewports in Paper Space can be activated by clicking on their boundaries and corresponding properties can thus be adjusted by right-clicking on the activated viewport. To avoid unnecessary clutter of lines, viewport boundaries can be removed by turning off viewport layer. That way, the drawing will appear more like a standard engineering drawing as shown in Figure 12.8. At this point, the drawing can be saved for printing or output on various formats as will be discussed in the next section, together with configuring printers and plotters.

12.5 CONFIGURING PRINTER OR PLOTTER

Although it is not absolutely necessary nowadays, drawings produced using CAD systems can be printed or plotted on paper using LaserJet printers or pen/ink plotters respectively, as discussed and shown in Figure 1.4. Most businesses nowadays have taken to use their electronic devices ranging from laptops, mobile phones and other such devices to view drawings in offices, construction sites and production plants. This has been made possible by

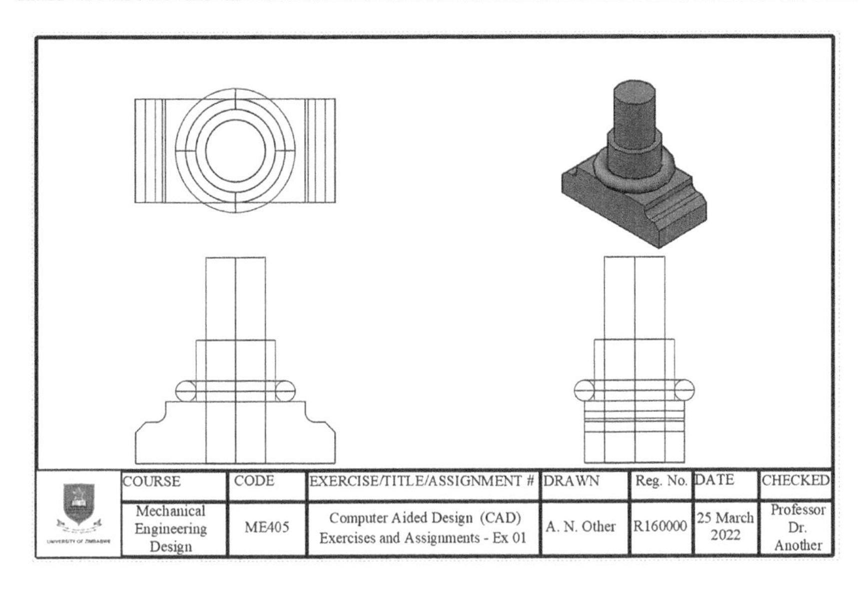

Figure 12.8 Cast iron base assembly organized for printing in paper space.

the various electronic formats that are now readily available, as will be discussed in this section. Most CAD systems including AutoCAD are now run under the Windows environment in which both plotters and printers can be configured through such platforms. Usually, each plotter or printer comes with installation files to enable it to be linked to any Windows operating software, in line with the operating manuals provided with the output device. The focus for this chapter will therefore be on the Print/Plot dialogue box.

12.6 PRINT/PLOT DIALOGUE BOX

The dialogue box for printing or plotting is similar to the second display in Figure 12.7 (right), which was activated using the **Page Setup Manager** and **Modify** but in the case of plotting or printing, the **File** pull-down menu followed by **Plot** is used to bring up an almost similar dialogue box as shown in Figure 12.9. This section will guide users on how to appropriately set up the various key parameters to ensure the proper output of the model or drawing on paper.

If a plotter is configured and linked to AutoCAD, it will appear among the list of possible printers or plotters. Usually, the name that appears there will be of the model of the connected printer or plotter and this can be selected as desired. If it is a new printer, it needs to be configured first in order for it to be recognized and listed among the available printers or plotters that can be used in AutoCAD. Just below the name of the output device, users can

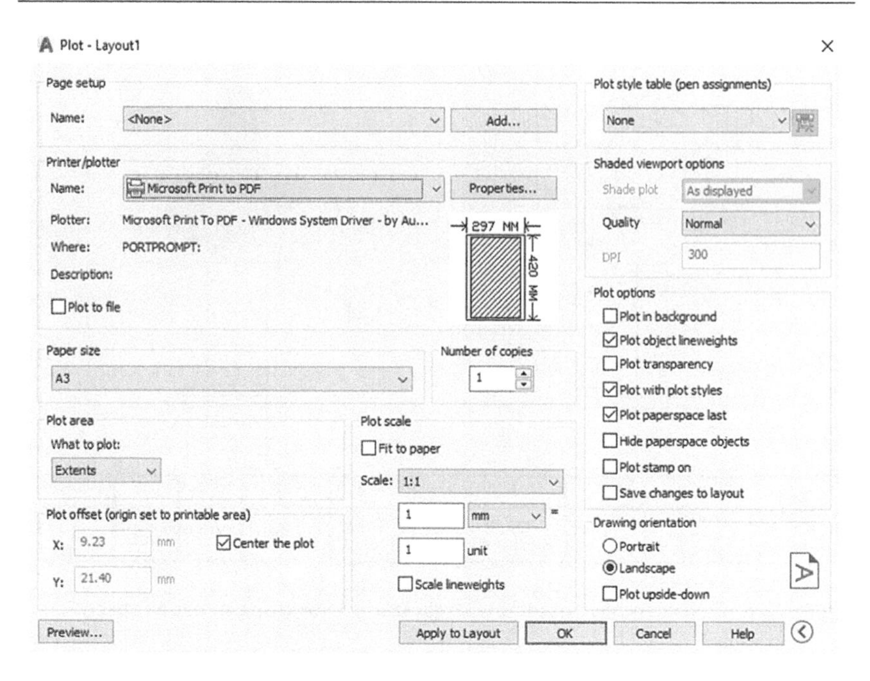

Figure 12.9 Print/plot dialogue box.

also specify the paper size, depending on what sizes the available output device can accommodate, alongside the number of copies required.

The plot area has a number of possibilities for what to print or plot. These are as follows:

1. Display: Prints/plots what is visible on the graphics window
2. Limits: Prints/plots what is within the set limits for the particular drawing
3. Extents: Prints/plots the drawing to the extents of the graphics window
4. Window: Users can select a window or portion of a drawing to print/plot

Regardless of the option selected, it is essential to have the correct display of what needs to be printed using the Preview Window of the Print/Plot dialogue box. If the drawing or model is not displayed correctly on the paper under Preview, adjustments can be made until the preview is displayed correctly. Such adjustments include checking Centre the Plot so that whatever needs to be printed or plotted is centrally displayed. The distance from the edge of the paper to the drawing boundary (plotter offset) can also be adjusted both on the x and y axis accordingly in order to get a desirable display. The drawing to be plotted or printed can also be scaled up or down depending on the size of the output paper. For example, if a model was developed using the A4 paper size limits (297 × 210), it can be scaled up by

2 to be plotted or printed on A3 paper (420 × 297). However, it is always advisable to use full scale right from the beginning, with the insight of what size the output media should be.

Instead of printing on paper, models and drawings can also be printed or plotted to electronic outputs such as plot files (Drawing.PLT). The plot file format or commonly referred to as PLT file, was developed by Autodesk as a vector-based plotter file, which can be saved on one device and printed or plotted on another device that may not be connected to the device where the model was developed. The use of PLT files guarantees accuracy and precision in production as all models and images are printed using lines and not dots as in the dot-matrix printers. PDF files are now the most commonly used output for models in electronic form, for use at construction sites or production plants.

Depending on the version of AutoCAD in use, some of the Print/Plot Dialogue Boxes are truncated to exclude some of the additional settings. However, these can be accessed by clicking the arrow to the bottom right of the dialogue box, if not visible. CTB is a file extension for colour settings used in AutoCAD. They are colour-dependent plot styles that are used to set how the 256 available colours in AutoCAD will appear when they are plotted or printed, including the lineweight assigned to each colour. When CTB styles are used, users will virtually be adapting some or all of the colours 0 through 255 to meet drawing office standards for plotting. Although CTB plot styles have been replaced by STB files, many companies continue to use CTB files due to their compatibility and familiarity.

Other additional settings include shaded plots, which specify what needs to be plotted, for example, 3D models with hidden lines removed or rendered etc. The quality of the printout specifies the clarity of the drawings. This can be adjusted up or down. However, it is costly in terms of charges for printing if models and drawings are of high resolution. Such high-resolution models or plot files are also large in terms of electronic storage. The settings for portrait and landscape allow users to adjust the orientation of the models and drawings so that they can be printed in the desired orientation.

12.6.1 Plot style/pen assignments

Depending on what output device or plotter is being used, this section is applicable primarily for colour plotters that make use of pens or pen turrets. This option allows users to specify the colour, pen number, linetype, speed and pen width. If the model was developed with different colours, then each colour should be assigned to a certain pen number. This should be followed by pulling out the pen turret from the plotter and the correct colours are then inserted in the corresponding pen positions in the turret. The default pen width is usually set at 0.254 but if a different pen width is required, this can also be appropriately set.

However, it is advisable to leave the pen width at the default AutoCAD value, and if necessary, polylines can be used to obtain any desired thickness of lines. The size option allows users to set the size of the output media, with labels such as User1, User2, User3, etc. where the width and height of the paper in use are specified. For example, an A3 paper would be (420 × 297) denoting the units that would have been used to set the width and height in millimetres.

12.6.2 Print/plot formats

As pointed out in earlier chapters, AutoCAD has several other formats available for outputting models and drawings. The standard and basic drawing format is the DWG which can be printed directly from model space or converted to PLT files for printing on a device that may not be connected to the one where the drawing was developed. However, such outputs are limited to what is on display in a particular viewport, if the viewports option is used. The most commonly used option for producing electronic drawings without the need for AutoCAD, is the PDF, which can be selected from among the available printers as Microsoft Print to PDF. Other commonly used formats that do not require AutoCAD for viewing include the DWF, which allows users to output their models and drawings in 3D Design Web Format for use on internet pages, the Windows Metafile (WMF) for use in Word, the Bitmap (BMP) image for use in Paintbrush and other associated applications as well as the Encapsulated Postscript File (EPS) for transferring vector artwork between applications.

Depending on where the drawings and models are required, the designers or users can make use of the several available facilities to organize and convert them for a particular purpose. If the drawings are required in a production plant or construction site, hard copies can be produced but more and more users now prefer to use different devices, in which case electronic versions are availed. For online advertising purposes, the DWF can be used.

12.7 CLOUD REVISIONS IN AUTOCAD

Revision clouds are a recent addition or improvement to AutoCAD, in which polylines in the form of arcs are used to enclose areas within an AutoCAD drawing in order to draw the attention of users to certain parts of the drawing. The command, REVCLOUD is used to create such areas in which designers may wish to draw attention of users for certain details or they may wish to highlight areas that have been revised since the last version of the particular drawing was issued. The sizes of the arcs or polylines used in revision clouds can be set to desirable sizes if the ones that are drawn by default are not big or small enough for the attention required.

In addition, the shape of the revision cloud can also be adjusted accordingly. All these settings are available after entering the REVCLOUD command and allow users to set Arc length/Object/Rectangular/Polygonal/Freehand, followed by the points enclosing the revision area. Users are encouraged to use a different layer or colour if such revisions are necessary, in order to clearly distinguish the arcs of the revision cloud from those of the main drawing. The object option allows users to first draw an object such as a rectangle, ellipse or circle and then use that as a base for enclosing the revision cloud. The other options are fairly straightforward as their titles suggest. Freehand allows the creation of a cloud revision which is done by free hand (holding down LMB) as it is drawn and released towards the end. As soon as the user reaches the starting point of the revision cloud, the curve will automatically terminate and a closed revision cloud will be made.

Created revision clouds can be modified, that is, expanded or reduced by using their grips, just like any other polyline or arc. More grips can be added to the revision cloud to allow for better control in case of adjustments. This can be done by changing the REVCLOUDGRIPS system variable to 0, resulting in polyline grips on every arc of the revision cloud whereas, setting the variable to 1 or ON will make grips only on the segments of the revision clouds.

Properties of the revision cloud such as the thickness of the arcs etc. can be modified by using the PEDIT command. Other modifications include arc length as well as the type of arc. The type of revision cloud required can be selected from rectangular, polygonal or freehand. The default style of a revision cloud is the series of simple arcs of standard thickness or width. However, users can also use more stylish and better-looking calligraphy-styled arcs. Figure 12.10 shows a typical layout drawing of the comminution

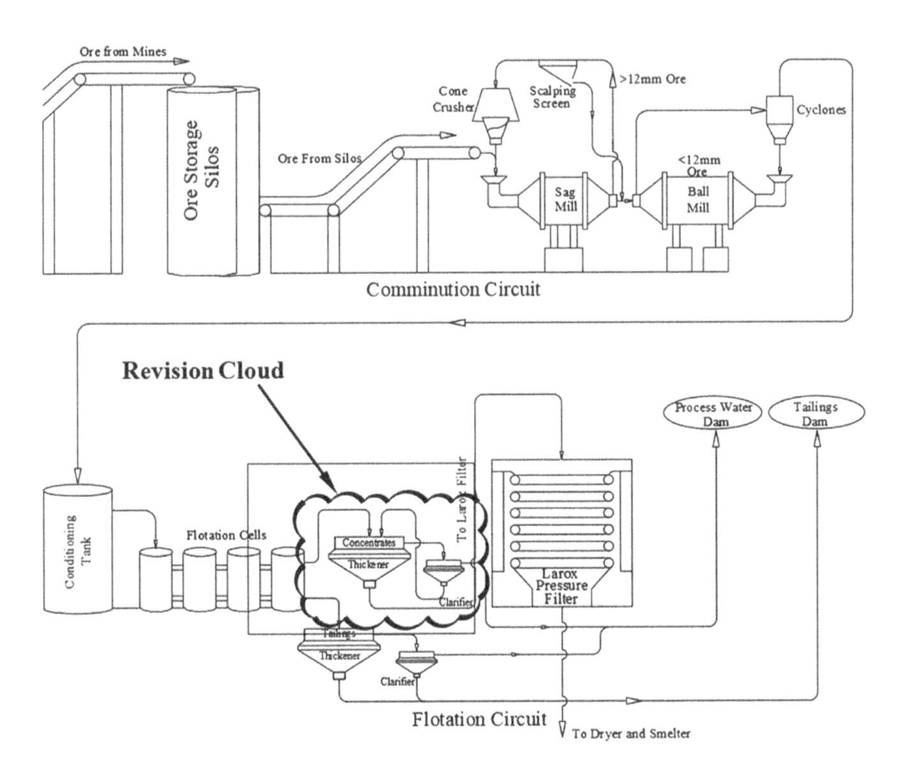

Figure 12.10 Platinum processing plant layout with revision cloud.

and flotation circuits of a platinum processing company. Assuming that the designers of this plant layout wanted to draw management or users to the area feeding concentrates into the thicknesser and clarifier within the flotation circuit, that area can be enclosed within the revision cloud, in a different colour and possibly also using the calligraphy style to additionally distinguish the revision cloud arcs from the other arcs of the plant layout as shown in Figure 12.10. Revision clouds are one of the important ways in which models and drawings are organized in today's environment.

12.8 SUMMARY

The ability to represent objects and systems using CAD systems in a presentable manner is critical for the realization of the benefits of investing in such systems. From Chapter 1 to Chapter 11, various skills were introduced on best practices in using CAD in order to realize a return on investment in CAD. These included drawings of simple primitives, combining and modifying them, dimensioning, layers, blocks and attributes as well as 3D modelling and customization. The ultimate purpose for all this would be to present detailed drawings and models in such a manner as to justify the transition from manual to computer-aided drawings. This would not only be useful to convince management to accept any new products that an organization may be considering embarking on but also to give credence and confidence to customers to buy the new products as potential buyers.

This chapter focused on how to guide users to organize and prepare their models for either printing or electronic presentations at construction sites or production plants by packaging them in such a manner as will be useful at such places. This also included the transition from Model Space (the usual drawing environment for AutoCAD) to Paper Space where the models are organized and packaged for sharing in different formats such as the commonly used PDF, Plot (PLT) files as well as Design Web Format (DWF). These formats also help to reduce the need for producing hard copies of drawings on paper, as was the case with traditional manual drawings, which were not only costly in terms of paper and handling but storage and consumables such as ink cartridges as well. Nevertheless, guidelines were also provided in this chapter to assist users on how best to configure their systems in preparation for the output of their models on paper or in electronic form. Another useful aspect for introducing revision clouds, a recent addition to the AutoCAD software was also included to assist users in drawing attention to certain parts of their models, all in a bid to maximize the facilities available in the CAD system.

12.9 REVIEW EXERCISES

1. There are principally two environments in which models and drawings can be developed and organized for output in AutoCAD, that is, Model Space and Paper Space. Clearly explain the difference between the two, outlining their applications, advantages and disadvantages.
2. Why are viewports necessary both in Model Space and Paper space in the development of models and drawings in AutoCAD? In addition, give practical situations and examples where such facilities are necessary and examples where it might be necessary to have viewports within other viewports.
3. Generally, when developing models using viewports, each viewport is enclosed within a given boundary. Such boundaries will create unnecessary clutter when preparing a drawing for printing. Explain how such boundaries can automatically be removed in Paper Space to leave only the contents of the viewport necessary for output on printers.
4. List and explain at least five formats in which models and drawings in AutoCAD can be converted and explain where each one of them is applicable in real-life situations such as production plants, construction sites or substations.
5. Explain the importance of revision clouds in AutoCAD, using one practical example.
6. List and explain, the five most critical parameters that should be properly set in the Print/Plot Dialogue Box in order to adequately organize models or drawings for output on plotters/printers or in electronic form.

Chapter 13

Further practical applications for CAD

Case studies

13.1 INTRODUCTION

While CAD systems can be used for various facilities in the creation and model representation of objects, they can also be used for practical applications in visualization and analysis. So far in the book, three case studies have been outlined along and within the appropriate chapters in which they reside. These case studies included the development of a CAD suite for the automatic generation of bills of materials for a motor vehicle manufacturing and assembling plant in Chapter 7, modelling a connecting rod for a diesel engine in Chapter 10 and customization of the CAD system and software at an electricity generation and distribution company in Chapter 11. These practical case studies served to demonstrate the importance and practical applications of CAD in engineering. However, there are numerous other cases where CAD systems have been implemented to improve the way engineering designers operate. This chapter will focus on an additional three case studies focusing on applications of AutoLISP in mechanism design and analysis, modelling and simulation of a dump truck tipping mechanism using EdenLISP and the digital inventory codification system for a tube and pipe manufacturing company, further demonstrating the importance of investing in CAD systems in this global era of competitiveness.

13.2 APPLICATIONS OF AUTOLISP IN MECHANISM DESIGN AND ANALYSIS

The data, modelling and analysis in this section are a summarized version of a case study for applications of AutoLISP, a derivative and interface programming language for AutoCAD, carried out to demonstrate its usefulness in mechanisms such as the four-bar linkage used in windscreen wipers for motor vehicles (Nyemba, 1999), which was subsequently published as a paper in the *Proceedings of the Zimbabwe Institution of Engineers*, a journal of the Institution of Engineers in Zimbabwe, from where more details can be obtained.

DOI: 10.1201/9781003288626-13

With the advent of the current Fourth Industrial Revolution and rapid changes in technology, the use of computers in engineering design is one of the most rapidly developing areas to cope with these changes. Even though in general, the world is currently immersed in the Fourth Industrial Revolution, many countries, especially in the industrializing world are still struggling to transition from the Third to the Fourth Industrial Revolution, when in fact, the rest of the industrialized world is already preparing for the Fifth Industrial Revolution (Nyemba, et al., 2021). This is very unfortunate, especially in this ever-increasing globalized world. The industrialized world may be producing technologies for consumption by the entire world but if other countries are lagging behind, this invariably results in inconsistencies and challenges, particularly for industrializing world, hence the move in tandem with the rest of the world. Such challenges have resulted in the need for good design tools to improve the interaction between the design and the designer.

Over the years, programming tools for CAD, including requisite software have increased both in numbers and capabilities. A few years ago, most CAD software was used primarily for producing detailed engineering drawings but over the years, engineers have demonstrated that the same tools can be used for a multitude of functions such as modelling and analysis. The fact that one day the computer will take over the bulk of the design process cannot be overruled as is already being observed in Artificial Intelligence (AI) and machine learning.

Modern industrial equipment would have been considered impossible a hundred years ago but in the 21st century, more and more such machines are running unattended because of their inherent ability in machine learning. With such technology dynamics and rate of technological development, today's impossibility may become an everyday commonplace in 20 years. However, to assert that computers will take over the whole design process may not quite be possible because human beings will still be necessary for programming, designing and instructing the machines. There is usually a misnomer that '*Computers are now asking humans to prove that they are not robots*' wherein the truth is that, for a machine to be able to ask such questions, a human being would have programmed it to do so.

This case study demonstrates one of the applications of AutoLISP, an interface programming language for AutoCAD, in mechanism design, analysis and simulation of a designed mechanism, in this case, a four-bar linkage mechanism, typically used in windscreen wipers for motor vehicles. The ability to use readily available software packages such as AutoCAD to perform this task can be useful to the engineer in a number of ways. These include checking for flaws or faults in a mechanism in order to rectify them before prototyping, manufacturing or implementation. Mechanism normally consists of a number of linkages, each of which has its own motion and the motion of one linkage has to be timed and synchronized with the other linkages to ensure smooth operation of the intended machine.

Any slight mistake or misalignment may result in clashes or undesirable consequences. It would therefore save a lot of time and money if these clashes are dealt with and solved at the conceptual stage of the design process before embarking on manufacture. The same designs or setups can be exported for finite element and stress analysis in other available CAD environments such as Solid Works, Ansys, Abaqus etc. This case study provides guidelines and procedures for using interface programming languages such as AutoLISP to model and simulate typical mechanisms for the purposes of analysing and optimizing designs.

13.2.1 Background to AutoLISP and case study

Most programming languages gained prominence in the mid- to late-1950s. Traditional and high-level languages such as Formula Translation (FORTRAN) and List Processing (LISP) were developed to overcome the problems of working in low-level machine code and assembly languages. However, these high-level languages were developed in two radically different ways, procedural and declarative (Mahmudova, 2018). FORTRAN was the first high-level procedural language to be developed followed by such languages as ALGOL, BASIC and PASCAL. Declarative languages, the first of which was LISP were also being developed during the same period.

LISP was developed by John McCarthy at the Massachusetts Institute of Technology in 1960 (Touretzky, 1990). Although LISP is mainly composed of symbolic expressions enclosed in a lot of parentheses, it highlights some of the delights of mathematical logic. LISP was developed initially for the study of AI, a subject also initiated by John McCarthy. Until the early 1980s when AutoCAD was introduced, LISP was mainly used for the study of AI on mainframe computers. AutoCAD then came along with the interface language, AutoLISP, a derivative of LISP and hence LISP became available on PCs. AutoLISP is an integral part of AutoCAD, written and distributed by Autodesk. It allows users and AutoCAD developers to write macro-programs and functions in a powerful high-level language that is well suited for graphics applications.

Unlike most procedural programming languages like FORTRAN and PASCAL, AutoLISP is a functional LISP language which is declarative. It however has the same basic structure as the other languages, which basically consists of:

- Defining functions
- Constructing lists
- Extracting parts of lists
- Using extracted values or variables to perform mathematical operations
- Displaying graphically in a CAD package what has been calculated
- Varying some of the variables and simulating the effects

In general, the development and use of AutoLISP on CAD and AutoCAD in particular, have been to assist the engineer in a number of different ways.

- The ability to get access to data
- Helps in routine or complex calculations
- Aids in visualizing, editing and hence optimization

13.2.2 Geometrical modelling of a slider-crank mechanism

The use of AutoLISP for mechanism design, simulation and analysis was carried out and demonstrated on a slider-crank or four-bar linkage mechanism, typical of which is used in windscreen wipers for motor vehicles. Linkage mechanisms are mechanical devices that are used for the mechanization and control of motion. Their principal application is in the conversion of an input motion into an output motion. Generally, there is either an input rotation or translation to an output rotation or translation. The slider-crank mechanism or four-bar linkage is generally an input rotation to an output rotation or vice-versa depending on the application, for example, the windscreen wiper is an input rotation to an output rotation or oscillation.

The basic and simplest mechanism or linkage system for most machines in this category is the four-bar chain, which consists of the crank, coupler, rocker and a fixed frame as shown in Figure 13.1. These four constitute four rigid links connected in the form of a quadrilateral arrangement with four pinpoints, hence the name four-bar chain or four-bar linkage. The angular positions of these links were calculated in relation to the angle of the crank to the horizontal that was assumed to be known or denoted as the starting point. The lengths of the coupler, rocker and frame were also based on the length of the crank. The equations derived, for the two particular cases of the crank geometry being above or below the frame, were used in coding the AutoLISP program for the eventual simulation of the mechanism.

The operation of such a mechanism as the four-bar chain is that there is an input rotation at the crank a through the coupler (b) to an output rotation (oscillation) at the rocker (c). The fixed link or frame (d) is where the mechanism is mounted. The crank rotates continuously and the output

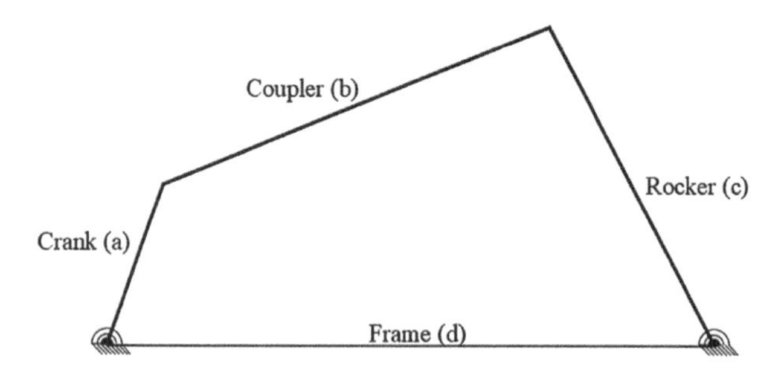

Figure 13.1 Four-bar linkage mechanism.

rocker oscillates. The first step in modelling such a mechanism was to derive the geometrical relationships so as to program the motions for simulation. Referring to Figure 13.1, the lengths for the crank, coupler, rocker and fixed frame were taken as a, b, c and d respectively. Such types of links as the four-bar chain can be divided into two classes that are used to develop the governing rules for such mechanisms (Vinogradov, 2020).

Class 1: For this class of mechanism both of the following criteria must be satisfied.

$$d < (b + c - a)$$
$$d > (|b - c| + a)$$
which leads to the range :
$$(|b - c| + a) < d < (b + c - a)$$

Class 2: This class comprises any other four-bar chain that is physically possible but fails to satisfy both of the criteria above. In this class of mechanisms, the links can only oscillate relative to one another. Thus, the focus of this case study was on the first case.

Another important consideration is that the linkage is physically impossible if one of the links has a length greater than the sum of the other three (Vinogradov, 2020).

For the determination of angular positions of each link relative to the horizontal (x-axis), it was important to note that there were two distinct cases in which the required angles would change; *Case 1*: the crank was above the line connecting the two fixed points (frame) and *Case 2*: the crank was below the line connecting the two fixed points(frame). Case 1 can be represented graphically as shown in Figure 13.2. For this case $0 \leq \theta_2 \leq \pi$. As is the norm

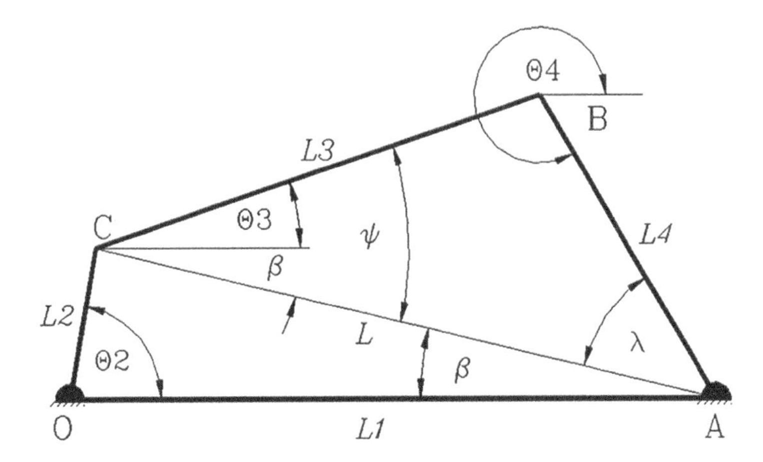

Figure 13.2 Four-bar linkage mechanism (crank above fixed frame).

in CAD systems, all angles were measured positive anticlockwise. θ_2 is the angle of crank to the horizontal and is assumed known as being the starting angle. All other angles in this case are as shown in Figure 13.2.

Considering triangle OAC in Figure 13.2,

$$\cos\theta_2 = \frac{L1^2 + L2^2 - L^2}{2L1L2}$$
$$\therefore L = \left(L1^2 + L2^2 - 2L1L2\cos\theta_2\right)^{\frac{1}{2}} \tag{13.1}$$

and

$$\frac{\sin\beta}{L2} = \frac{\sin\theta_2}{L}$$
$$\therefore \beta = \sin^{-1}\left(\frac{L2}{L}\sin\theta_2\right) \tag{13.2}$$

However, to leave b defined as the *inverse sine* of an angle would not have been very helpful since there is no built-in AutoLISP function for *inverse sine*. It was therefore necessary to define the *inverse tan* of the angle since there is a built-in AutoLISP function (**atan**).

From $\tan x = \dfrac{\sin x}{\cos x}$ and $\cos^2 x + \sin^2 x = 1$,

b can be expressed as: $\beta = \text{atan}\left(\dfrac{\sin\beta}{\cos\beta}\right)$

and since $\sin\beta = \dfrac{L2}{L}\sin\theta_2$, $\cos\beta = \left(1 - \left(\dfrac{L2}{L}\sin\theta_2\right)^2\right)^{\frac{1}{2}}$

Therefore:

$$\beta = \text{atan}\left[\frac{\frac{L2}{L}\sin\theta_2}{\left(1 - \left(\frac{L2}{L}\sin\theta_2\right)^2\right)^{\frac{1}{2}}}\right] \tag{13.3}$$

Considering triangle ABC

$$\cos\varphi = \frac{L^2 + L3^2 - L4^2}{2LL3}$$
$$\therefore \varphi = \cos^{-1}\left(\frac{L^2 + L3^2 - L4^2}{2LL3}\right)$$

Similarly,

$$\cos\varphi = \frac{L^2 + L3^2 - L4^2}{2LL3}$$
$$\therefore \varphi = \cos^{-1}\left(\frac{L^2 + L3^2 - L4^2}{2LL3}\right) \qquad (13.4)$$

θ_3 and θ_4 depend on whether its *case 1* or *case 2*.
From Figure 13.2 (*case 1*) and considering triangle ABC.

$$\frac{\sin\lambda}{L3} = \frac{\sin\varphi}{L4}$$
$$\therefore \lambda = \sin^{-1}\left(\frac{L3}{L4}\sin\varphi\right)$$
$$\therefore \lambda = \operatorname{atan}\left[\frac{\frac{L3}{L4}\sin\varphi}{\left(1-\left(\frac{L3}{L4}\sin\varphi\right)^2\right)^{\frac{1}{2}}}\right]$$
$$\therefore \theta 3 = \varphi - \beta$$
$$\text{and}\,\theta 4 = 2\pi - \lambda - \beta$$

Case 2 can be represented graphically as shown in Figure 13.3 where the crank is below the fixed frame. The variables defined in *case 1* were the same for this case except for θ_3 and θ_4.

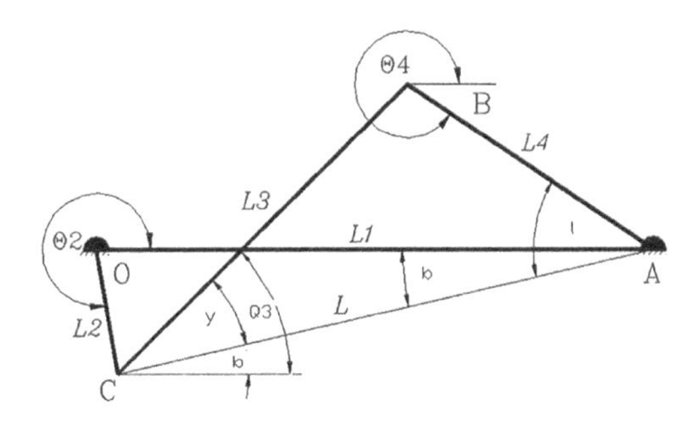

Figure 13.3 Four-bar linkage mechanism (crank below fixed frame).

From Figure 13.3 and considering triangle ABC:

$$\frac{\sin\lambda}{L3} = \frac{\sin\varphi}{L4}$$

$$\therefore \lambda = \sin^{-1}\left(\frac{L3}{L4}\sin\varphi\right)$$

$$\therefore \lambda = \operatorname{atan}\left[\frac{\frac{L3}{L4}\sin\varphi}{\left(1-\left(\frac{L3}{L4}\sin\varphi\right)^{2}\right)^{\frac{1}{2}}}\right]$$

$$\therefore \theta_3 = \varphi + \beta$$

$$\text{and}\,\theta_4 = 2\pi - \lambda + \beta$$

13.2.3 AutoLISP programming for simulation and analysis

The sequence of steps that were used to derive and code the AutoLISP program to model and simulate the four-bar linkage mechanism was based on the calculations in the previous section. The AutoLISP program is in Appendix A13, laid out on in the same sequence as outlined below. Where necessary, additional statements or explanations were provided along with the program code. These explanations are found on every line that is preceded by (;). The general sequence for the procedure adopted in coding the AutoLISP program can be summarized as follows:

1. Obtain the lengths of links from the user and also define a starting point, say $\theta_2 = 0.0$ (or any other suitable value)
2. Calculate $L, \beta, \varphi, \lambda, \theta_3, \text{and}\, \theta_4$
3. Define coordinates for frame, crank, coupler, rocker
4. Display the four links, frame, crank, coupler and rocker on the AutoCAD screen
5. Increment θ_2 by 0.1 (or any other suitable value)
6. Delete the current display to start the simulation
7. Recalculate θ_3 and θ_4 depending on whether the crank is above or below the horizontal and repeat the above procedure to continue with the simulation
8. End.

The program contained one subroutine at the beginning and this was used to get the user's input, that is, the geometry of the four-bar chain being studied. The values that the user enters are tested in this subroutine. If they satisfy the conditions for *Class 1*, as spelt out in Section 13.2.2, then the program proceeded and executed the rest of the statements. However, if the values entered did not agree with the *Class 1* criteria, then an error would be generated and the user was prompted to re-enter the geometry of links. These conditions are spelt out in the program listing in Appendix A13. The loops created by recalculating and redefined the various angles, coupled with erasing and redrawing the model, resulted in the animation or simulation of the four-bar chain.

13.2.4 Results and implications

The great diversity of design activity has often been noted with all designs represented as compromising two broad functions, that is, decision making and information processing. The former includes the development of the most satisfactory solution to a problem, that is, optimization which invariably leads to iterations and sometimes visualization (display). Therefore, there is a need for iterative methods and graphics facilities, especially interactive graphics when the frequency of decision-making is high and time compression is important.

Information processing is another aspect and vital function of design. This can also be aided by the use of interactive graphics. There have been problems in developing the types of information processing that are needed for design purposes. Economic pressures encourage the development of an overall cost-effective system approach. CAD has been depicted as comprising two distinct disciplines, graphics and modelling (Vinodh et al., 2010). Considering the illustrated example of the four-bar chain, the component drawings can be displayed graphically. Modelling, on the other hand, consists of the use of the program accessing a database to compute the position of links when one of them is varied in order to achieve simulation and visualization of the motions on a computer screen.

In general, interactive graphics are used in the design environment to optimize the design of parts, components, assemblies, etc., before the manufacturing phase commences. The aim would be to ensure that a product is well designed before the costly exercise to manufacture it. This can avoid expensive work associated with such activities as prototyping and introducing modifications to items after manufacturing has begun.

Linkage analysis, for example, was intended for the design of mechanical linkages and to investigate their properties. It is possible to enter the input data in a matter of a few minutes and the user can perform a complete geometrical and motion analysis in even less time. Parametric analysis can also be executed for the different forms of engineering design problems, for example, in structural design and mechanisms of machines in general. With time, engineering structures generally fail, depending on the materials used, the kinds of loads in operation and conditions of support. They can be modelled mathematically and the critical loading conditions and factors of safety calculated in order to optimize and improve the life cycle for components.

With rapid changes in technology, there has been a corresponding increase in either new design software or the improvement of the existing ones.

While many companies in industrializing countries continue to invest in CAD systems, this has been largely for the purposes of drafting only but more and more are turning to fully utilize the CAD systems for other purposes such as mechanism design, analysis and parametric modelling. This has been largely due to global competition and the need to keep abreast with the dynamics of the rapid changes in technology as well as increased awareness of the benefits of CAD software. Unfortunately, very few if any at all, even in the industrialized world, make use of AutoLISP programming, mainly because of a lack of the appropriate skills. However, it is hoped that such case studies will be useful to further raise and aware and encourage AutoCAD users to make use of such benefits that come as part and parcel of all versions of the AutoCAD software.

The processing files written in AutoLISP can be modified and customized to suit a user's specific needs, thus presenting the user with more control over the design program. As the saying goes, 'Using one's own creation is much simpler than somebody else's', this also rings true in the use of AutoLISP programming where a lot of control is required to develop sound designs, thus making engineering design easier even for those with little or no practical experience.

Analytical graphics is a vital aspect for undergraduate engineering students and this can be accomplished through the use of CAD systems, such as in the case of the four-bar chain that has been demonstrated. The added use of simulation and animation of such analytical graphics make visualization and understanding of engineering principles easier and more enjoyable. On the other hand, improved interactions can also be achieved by using AutoLISP for geometrical modelling and analysis, based on mathematical models, the base on which engineering problems are solved using iterations and parametrics. This ultimately helps organizations and engineering designers in the decision-making process of design.

13.3 MODELLING AND SIMULATION USING EDENLISP

The data, modelling and analysis in this section are a summarized version of a case study for modelling and simulation using EdenLISP, a derivative of the AutoLISP interface programming language for AutoCAD, carried out to demonstrate its usefulness in typical machine design of a dumping truck tipping mechanism (Nyemba, 2013), which was subsequently published as a paper in the Journal of Science, Engineering and Technology of the Zimbabwe Institution of Engineers, from where more details can be obtained.

Theinterface programming language for AutoCAD, AutoLISP is a derivative of the high-level LISP programming language and allows users to develop their own macro-programs coded in LISP for graphical applications using functional programming. It also allows users to develop their own programming environments. As was alluded to in the previous case study, AutoLISP requires some skills and experience in programming as many may find the use of parentheses a bit annoying. These cannot be avoided as they are the controlling factor in all LISP programs. In order to ease this challenge, a simpler programming environment was developed by Allan Cartwright at the University of Warwick in 1994. This was coined the Evaluator for Definitive Notations coded using AutoLISP (EdenLISP). EdenLISP is a simplified programming environment that makes use of declarative notations and not parentheses as in AutoLISP, even though it was coded in AutoLISP (Cartwright, 1994). This simplification makes it more user-friendly and easily accessible by engineering designers.

The application and use of AutoLISP for mechanism design and analysis were demonstrated in the previous case study for the four-bar linkage mechanism. The same principles were employed in this case study, except that EdenLISP was used on the tipping mechanism of a dump truck. Mechanical engineering designs revolve around machines and their mechanisms, forming the basis on which such analytical programs are developed for geometrical modelling, analysis and simulation to detect and predict failures in such equipment.

The operations and strengths of components in machinery and equipment are critical for mechanical engineering designers, hence their quest to develop systems and programs to study and analyse these in order to optimize them before prototyping and manufacturing. As a result, software developers as well as engineering designers have developed several routines in many different CAD packages to enhance engineering design in the product development cycle. As illustrated in the previous case study on the four-bar chain, this case study went further to another practical application in the form of the tipping mechanism for dump trucks that are commonly used by local government authorities in the collection and disposal of waste. The mechanism components were initially broken down and geometrically modelled into simple lines or links, programmed, simulated and analysed using EdenLISP.

The geometrically modelled components of the tipping mechanism were studied and observed practically on such dump trucks in order to establish and time the various motions to enable input to the EdenLISP compiler and eventual simulation and analysis of the mechanism. Such analyses empower mechanical engineering designers to identify bottlenecks in machine operations as well as areas of interference in operation, thus adequately designing machines that are not only safe but sound and long-lasting.

Geometrical modelling in the context of this case study entailed the construction and investigation of a machine in the form of a dump truck mechanism in order to discover characteristics of a fully developed version of the same machine. Complex as the dump truck may seem to a layman or engineering designer, it can be simplified by representing the moving and connected components in CAD using simple lines and other primitives, to enable visualization and appreciation of its characteristics and behaviour in operation in real life, through animation or simulation of the modelled links. In the 21st century and in line with the Fourth Industrial Revolution, engineering designers are sometimes regarded as 'modern' because of their ability to model using CAD as opposed to the traditional craftsmen.

Traditionally, improvements in product designs were done by effecting a series of small evolutional steps, followed by physical prototyping and testing. This was costly in terms of materials for the prototypes, hence the new thrust to invest in CAD in order to carry out all these iterations on the computer before prototyping. In recent days, models are now commonplace and enable engineering designers to tailor-make products to customer satisfaction before embarking on full-scale production. This would however require some knowledge and incite on the strength, operating characteristics and timing of motions for machine components as well as performance prediction of complex systems.

Although there are many other avenues that have been adopted for modelling such as anthropometrics where models are represented by statistical tables and graphs in order to aid the design process, this is generally not adequate for engineering designers, hence the use of sketches and drawings to adequately represent objects and machines. In this case, visualization is key and an important aspect of developing sound designs, thus the need for programs to display and visualize the links in motion in order to optimize their functioning.

13.3.1 Background to EdenLISP and case study

The development of high-level programming languages in the 1960's emerged as either procedural or functional routes. Within the two categories, also emerged several different environments such as FORTRAN, Pascal, Basic etc. in the former and several dialects from the general LISP such as Common LISP, MacLISP, AutoLISP and more recently EdenLISP. EdenLISP was derived using AutoLISP to simplify the complications encountered in dealing with so many parentheses as in the AutoLISP. This is partly because in general functional programming languages are not as straightforward and easy to understand and use as most procedural programming languages such as BASIC FORTRAN and C which are mostly used for mathematical modelling and not graphical applications.

EdenLISP was developed by combining key aspects of functional programming languages such as LISP through which AutoLISP was used for coding, as well as procedural programming languages aspects of Evaluator for Definitive Notations (EDEN). The programming environment for EdenLISP and the syntax contain English-like expressions but still maintain the graphical functionality of AutoLISP, hence a combination of user-friendliness and components necessary for visualization engineering design (Cartwright, 1994).

13.3.2 Geometrical modelling of the dump truck mechanism

Several typical engineering mechanisms were used to test the efficacy of EdenLISP during its development and eventual launch in the 1990s. One of those mechanisms used for such a test was the bell-crank mechanism for Wallpact Limited, a professional engineering company that was based in Rugby, England. The bell-crank mechanism was successfully modelled and simulated using EdenLISP. The same principles were employed in modelling the dump truck mechanism for a company in Harare, that used dump trucks for waste collection and disposal. Figure 13.4 shows a typical schematic for a dump truck, which consists of a tray resting on a bell-crank, with one arm connected to a hydraulic cylinder and the other to the tray.

A lever was connected to the hinge of the bell-crank. This mechanism and its operation were very similar to that which was studied at Wallpact. The technique of modelling such a mechanism in CAD was one for deciding how to partition the components of the bell-crank mechanism into suitable discrete components in such a way that sufficient flexibility is retained to explore the behaviour of the resulting model in a wide enough range of situations. In modelling such a machine, the focus was on the tipping mechanism and the tray, although the totality of the machine including the truck horse was not ignored. Figure 13.5 shows the simplified CAD geometric model after extracting and drafting the mechanism, with the following link configurations.

2-3-4: Bell-crank
6-7-8-9: Tray
4-6: Hydraulic cylinder and
1-5: Frame between the 2 wheels of the dump truck

The operation of such a mechanism was that assuming the tray was loaded with waste and ready to be dumped, the hydraulic cylinder 4–6 was

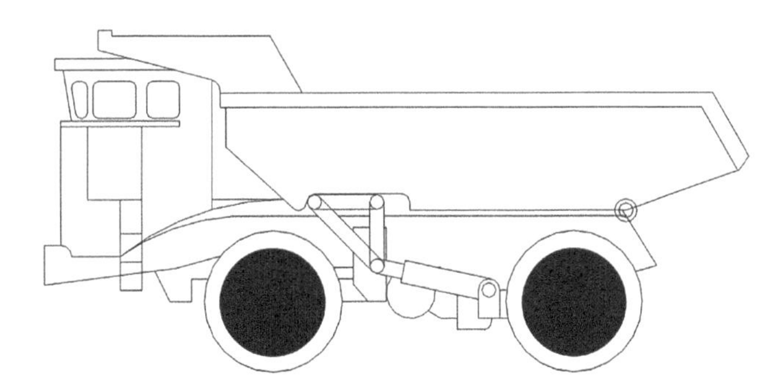

Figure 13.4 Typical dump truck.

(*Source:* Nyemba, 2013)

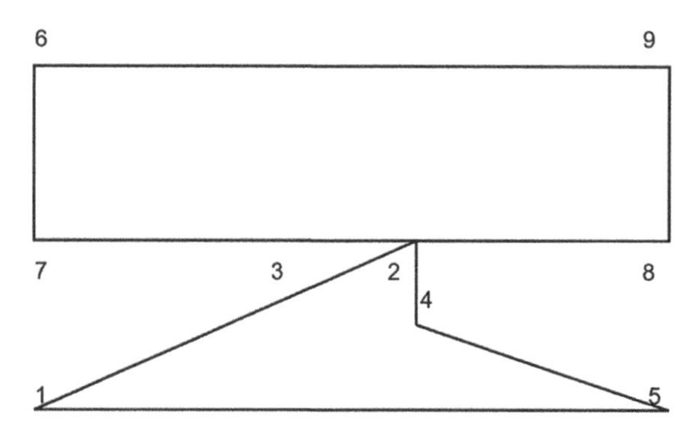

Figure 13.5 Geometric model for dump truck tipping mechanism.

(*Source:* Nyemba, 2013)

activated and as the ram opened, it pushed the lower arm of the bell-crank (4) that rotated about the hinge (2). In turn, the upper arm (3) pushed the container and tipped towards the right. The lower right corner of container (8) was fixed but rotated about that point.

The engineering design of such a mechanism revolved around geometrical modelling and timing of motions in such a way that the links were correctly positioned at each instant that the ram pushed the bell-crank until the lower arm was level with the lever, at which point, the lever and the arm were concurrently pushed. Studying and observing the mechanism further, it was apparent that the ram executed both translation (pushing the bell-crank) and rotation (oscillating about point 5). These motions resulted in two arcs, one centred at 5 and the other at 2, intersecting at 4 as the ram opened up. Points 1, 2, 5 and 7 were initially fixed and could be defined by the respective (x, y) coordinates. The other points depended on these four. However, when the length 4–5 changed, so did the points 3, 4, 6, 7, 8 and 9, as shown in the dashed profile in Figure 13.6.

13.3.3 Mathematical modelling of the dump truck mechanism

The information (specifications) provided and also confirmed by measurement for the dump truck tipping mechanism were as follows: Lever length: 2,800 mm, Arm length for bell-crank: 500 mm, Tray dimensions: 4400 × 1000 mm and Angle of lever to horizontal: 23°.

The first step in developing the mathematical model was to establish the coordinates of the point (4) as this was the intersection of the two circles as the ram opened up. Point (5) was selected as the origin of the mechanism, purely for simplifying the model. Figure 13.7 shows the intersection of the arc formed by the ram as it opens and the circle formed by the bell-crank, thus the two intersections of circles.

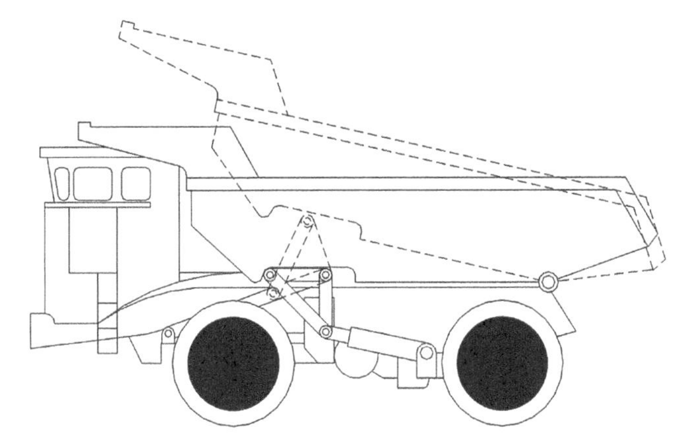

Figure 13.6 Dump truck in tipping position (dashed profile).

(*Source*: Nyemba 2013)

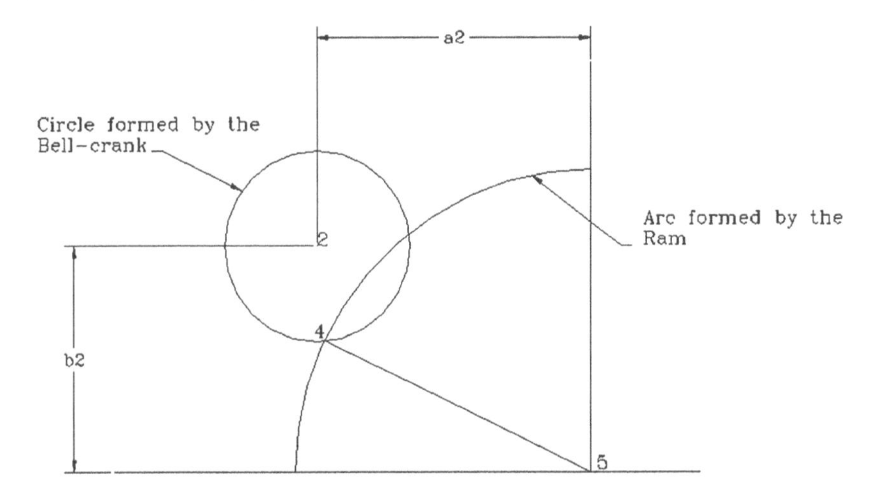

Figure 13.7 Intersection of circles formed by ram and bell-crank.

(*Source:* Nyemba 2013)

The intersection of two circles is governed by Equations (13.5) and (13.6)

$$x^2 + y^2 - 2a_1x - 2b_1y + c_1 = 0 \tag{13.5}$$

$$x^2 + y^2 - 2a_2x - 2b_2y + c_2 = 0 \tag{13.6}$$

Where (a_1,b_1) is the centre for circle 1 and (a_2,b_2) is the centre for circle 2. Hence:

$$c_1 = a_1^2 + b_1^2 - r_1^2 \tag{13.7}$$

$$c_2 = a_2^2 + b_2^2 - r_2^2 \tag{13.8}$$

Where r_1 and r_2 are the respective radii of the two circles. Since point 5 has been chosen as the origin or (0, 0) then $a_1 = b_1 = 0$ and Equation (13.5) reduces to:

$$x^2 + y^2 = -c_1 \tag{13.9}$$

and thus,

$$y^2 = -c_1 - x^2 \text{ and } y = \sqrt{r_1^2 - x^2} \tag{13.10}$$

Substituting $-c_1$ in Equation (13.6) gives

$$y = \frac{c_2 - c_1 - 2a_2x}{2b_2} \tag{13.11}$$

and Equations (13.10) and (13.11) give

$$\frac{(c_2 - c_1 - 2a_2x)^2}{4b_2^2} = r_1 - x^2 \tag{13.12}$$

Equation (13.12) can be further simplified and reduced to a quadratic equation in x

$$(4a_2^2 + 4b_2^2)x^2 + (4a_2c_1 - 4a_2c_2)x + (c_2^2 + c_1^2 + 4b_2^2c_1 - 2c_1c_2) = 0 \tag{13.13}$$

the solution of which can be derived from solving a quadratic equation thus,

$$x = \frac{-B \pm \sqrt{B^2 - 4AC}}{2A}$$

Looking back at Figure 13.7, it should be noticed that there are two intersection points and the one required (point 4) would be the one with the lowest value (more negative) of x, hence:

$$x = \frac{-B - \sqrt{B^2 - 4AC}}{2A} \tag{13.14}$$

and from Equation (13.11),

$$y = \frac{c_2 - c_1 - 2a_2x}{2b_2} \tag{13.15}$$

The (x,y) coordinates for point (4) are therefore derived from Equations (13.14) and (13.15) as the ram opens outwards. At the starting point, the end of the ram would be at a distance equivalent to the stroke from the base of the cylinder. Therefore:

$$\text{stroke} = \sqrt{{a_2}^2 + (b_2 - \text{armlength})^2} \tag{13.16}$$

Using the specifications provided and confirmed by measurements of the dump truck tipping mechanism, the stroke was found to be 1917 mm at the initial position (Nyemba, 2013). During motions when the lower arm of the bell-crank came in line with the lever, both linkages traversed outwards together. Therefore, from this point, apart from parametrically varying the length of the ram length or stroke, the angle of inclination of the lever would also vary. This resulted in the variation of the relationship of point (3) from point (4) where conditions were included in the EdenLISP design program to utilize a different formula for deriving the position of point (3) which invariably affected points (6) to (9).

13.3.4 Simulation of the mechanism using EdenLISP

As was done for the four-bar chain, simulation or animation of the mechanism was achieved by programming routines and loops based on the geometrical and mathematical models derived in previous sections above except that these were coded in EdenLISP as shown in Appendix B13. The links were initially displayed on the CAD screen using the stroke of 1917 mm and an angle of inclination of the lever of 23°. To achieve the animation, this was followed by erasing the first display, then varying the stroke and angle of inclination based on the formula and equations program and then displaying the model again.

These routines were repeated by automating the program using loops and variating the angle of inclination and stroke with every display, resulting in the simulation. Although the program may be long and seemingly difficult to execute, the calculations and determination of new positions as the simulation progressed, as well as erasing and redisplaying the model happened so fast that it was hardly noticeable that something was happening behind the scenes. Instead, the animations on the AutoCAD screen appeared purely and almost like a proper motion of a machine in operation. The simulation helped to identify areas that needed adjustment such as increasing the stroke so as to tip the tray further and empty all its contents.

The program in Appendix A2.2 was also formulated in such a way as to stop the simulation when the tray had completely emptied all the waste in it (fully tipped). Alternatively, the simulation could also be stopped at any given time by the user by simply pressing the Esc button. Improvements were also made to the model by incorporating different colours for the links in order to adequately visualize the mechanism and simulation.

13.3.5 Results and implications

The EdenLISP program listed in Appendix B13 was based on the derivation of positions of the links based on the geometrical and mathematical models in Sections 13.3.2 and 13.3.3. This program listing is evidently simpler to follow and thus user-friendly as compared to the AutoLISP program for case study 1, as they are expressed in plain statements but with very minimal parentheses typical of common LISP. However, the program syntax and structure are almost similar to that for AutoLISP which consists of firstly declaring whether or not variables such as points and lengths were absolute distances, reals or integers.

The variables were then defined by using the formulae derived in Sections 13.3.2 and 13.3.3 to determine the x and y coordinates of the points from one stage of the simulation to the next. The dependent variables (x, y) coordinates in this mechanism design were utilized before they were defined, which is a useful aspect of the EdenLISP programming routine. The values for these coordinates were used to determine the coordinates for point (4) in particular and the rest of the points derived from this point (4) using the polar coordinate facility in AutoCAD. The conditions for enabling the lever and bell-crank to move together once they were on the same level were also incorporated into the EdenLISP program. The model geometry was initially displayed with a stroke of 1917 mm followed by calculating the position of the other links, erasing them, recalculating at a new angle of inclination and then redisplayed on the graphics screen to achieve the animation effect of the mechanism.

The analysis of such a mechanism included adjustments of the links in order to achieve full tipping of the tray. However, it was also programmed in such a way that it does not necessarily need to tip over completely unless the tray was full, to push the ram to full length and tip over the waste. All these changes were possible and achievable because of increased computer power and the rate at which computations can be done while graphically displaying the model on the screen. In some instances, the incremental angle of inclination had to be lowered in order to reduce the speed of the animation in the cases of computer hardware that had high frequencies such as 1GHz and above, which is more or less the norm in today's high-speed computing.

This example of mechanism design analysis demonstrated the capabilities of CAD systems to carry out parametric analysis for different forms of engineering design problems such as structural design and integrity, dynamics and motion analysis of machine components. Quite often, engineering components fail due to improper movements while in operation, the kind of loads that are subjected to the components as well as the materials used in their manufacture. All of these can be assessed and optimized at the conceptual stage of the design process and corrected before embarking on costly manufacture of prototypes or the actual machines. Critical loading conditions can also be determined while factors of safety are kept in check to ensure long life spans for the machines.

The development of a new programming environment in the form of EdenLISP, made use of both aspects of functional and procedural programming languages, resulting in a user-friendly and easy-to-use software variation of the LISP programming language. The use of EdenLISP in this case study was another demonstration of the diverse facilities available in AutoCAD that are often underutilized. As with AutoLISP, EdenLISP also improves interaction in geometrical and general engineering design and analysis as well as optimization through parametrics, modelling and simulation to improve visualization.

13.4 DIGITAL INVENTORY CODIFICATION SYSTEM

The data, modelling and analysis in this section are a summarized version of a case study for the development of a 10-digit inventory codification system for a tube and pipe manufacturing company based on CAD systems and recent advances in technology and the desire to improve inventory management in manufacturing companies (Nyemba & Mbohwa, 2017). This was presented as a conference paper at the Fourth Global Conference on Sustainable Manufacturing in Stellenbosch, South Africa in 2016 and subsequently published as a paper in the Journal, *Procedia Manufacturing*, from where more details can be obtained.

In this global era of competition and the Fourth Industrial Revolution, companies continue to gear themselves to remain on the market. This includes ensuring that they remain competitive as they move in tandem with rapid changes in technology. The use of CAD in the drawing and design office would not be beneficial unless the whole organization is computerized in the same manner. Some of the case studies outlined in this book so far have revealed the need to link graphic information developed in the design office to non-graphic information used for planning and scheduling in the production plant, purchasing and sales departments. A complete revamp and computerization of an organization should invariably include inventory management, in terms of managing stocks, whether it is raw materials for use in production or finished and processed products for sale.

The integration of the various functional departments within an organization is a vital cog in the realization of the benefits of fully computerizing. The absence of proper inventory control systems normally results in high stock-outs, leading to possible losses on the share market. This case study was carried out at a company that specializes in the manufacture of tubes and pipes for various sectors of the economy, including mainly agriculture and mining. The focus was on the technical and engineering services required to maintain smooth production with a robust inventory control system to ensure efficient utilization of resources.

13.4.1 Background to case study and inventory control

The case study revolved around carrying out an As-Is-Analysis outlining their operations, limitations and production process flows, in conjunction with, and with particular focus on, their inflows (raw materials and parts) and outflows (processed products for sale). This led to the establishment of an inventory audit, a major input to the eventual development of a 10-digit inventory codification system that was designed as a generic tool with the capabilities of making decisions on the procurement of spare parts, in line with their CAD system in the technical services department (Nyemba & Mbohwa, 2017). Although the case study revealed that it would initially be costly to implement, reductions in inventory could well be realized through the removal of slow-moving and obsolete stocks, coupled with improvements in record keeping and accountability.

Traditionally, the scope for inventory control has been limited to production with a very remote link to engineering and technical services. However, there has been a paradigm shift in recent years for incorporating it in engineering management and technical services of companies, as these provide direct support to production (Liu & Ridgway, 1995). The penultimate aim in engineering control systems would be to realize a good return on the products and services provided by a company, through the maintenance and design of manufacturing equipment at a low cost. Companies without proper inventory control systems risk losses that are brought on by deterioration in inventory, stock-outs, high capital investments for inventory, difficulties in tracking inventory as well as high maintenance costs, all of which invariably and negatively affect their output (Kutzner & Kiesmüller, 2013). The case study company manufactures a wide range of steel tubes and pipes and other related products using different machine tools such as presses, welders, extruders and drawing machines, supported by their engineering and technical services, in terms of designs and maintenance of all production equipment.

One of the major challenges established at the beginning of the case study was that the company had huge amounts of stock of their inventory in finished goods (outflows), which was initially deduced as an indication that what they were producing was not exactly what the customers wanted, but of course, they continued to order raw materials and kept on piling their stocks. This was one of the major motivations to carry out this work for inventory profiling. It was also observed that the procurement and planning for spare parts prior to maintenance of machinery were haphazardly carried out, resulting in stock-outs. The maintenance records were not very clear in terms of location of the parts repaired or replaced, let alone difficult to assess the behaviour of the machines over time, sometimes leading to unplanned maintenance. In the absence of machine details including serial numbers, modifications on manufacturing equipment could thus not be tracked. Having established the company operations and its challenges, the case study then focused on the development of an inventory control and management system for the engineering and technical services function to ensure effective utilization of resources and management of inventory.

13.4.2 Inventory control systems and behaviour

There are typically four inventory control systems that companies can use for the effective management and control of inventory (Panneerselvam, 2012). These include:

Single bin system, also known as the P system, where the shelf or bin is filled up periodically,

Two-bin system, consisting of two compartments where the front compartment contains materials issued and the back compartment is sealed. When material in the front compartment is finished, the back compartment is opened for use and an order is placed (Q system).

Card system usually contains one card for each inventory item kept and has decision rules for either the P or Q system.

Computerized system records a computer-readable storage which is maintained for each item.

According to Panneerselvam (2012), considerations should be given to sizes of inventory, ordering parts or spares, and capacity-related costs when making any decisions with regard to inventory management. An essential determinant for an effective inventory is an accurate forecast of demand which can either be independent or dependent. Distorted records or information from one end of the supply chain usually result in uncertainties. This invariably leads to excessive inventory, poor customer service, wrong capacity plans and high costs, a common result of the bullwhip effect (van Horenbeek et al., 2013). This usually results from slight to moderate demand uncertainties and variabilities as well as stockpiling and excessive and idle inventory if each supply chain member makes ordering and inventory decisions without consulting the others. However, due to supply and demand, inventory has to be balanced continuously and should not be reduced if there is likely to be a negative effect on customer service.

Forecasting in inventory management is usually used to determine the value of a variable at some point in time, based on the fact that if the future can be predicted, then the company's practice of spending on spare parts can be adjusted accordingly. Forecasting can also be carried out in the short term when decisions are made on inventory control and production planning in medium term when tactical decisions in line with company strategic objectives are made. In doing so, forecasting can be qualitative, in the absence of relevant past data as in the case of the tube and manufacturing company, or quantitative methods such as analytical or statistical analysis, regression, moving averages or exponential smoothing can be used to establish the value of inventory. Replacement of assets or inventory frequently occurs when new parts come in with new technology, especially in this era of rapid changes in technology, which may lead to rendering existing spare parts in stock, obsolete. Under such circumstances, decisions to replenish spare parts can thus become complex and hence require appropriate tools and measures to be taken without affecting the smooth flow of production and business.

13.4.3 Data collection and analysis

The research carried out, based on the scant records, categorized three classes of inventory: *Slow moving* constituting 19%, inventory on hand and in excess of a year's requirements, or inventory on hand that had not moved in six months; *obsolete inventory* constituted 24% inclusive of all inventory items purchased or produced but had not been moved into production or sold within 12 months and 57% constituted the rest, which could not be classified as either slow-moving or obsolete. It was further established through interviews with personnel that some of the causes of obsolescence and slow-moving inventory included inaccurate forecasting or irregular supply of parts, which required huge safety stocks to cover uncertainties, large batch sizes and long machine setups resulting in high levels of work in progress, management desire to keep workers occupied, design and specification modifications in products as well as changes in production methods.

Removal of such excess inventory on the other hand meant a reduction in the unnecessary holding costs and an increase in the investment by taking care of the opportunity costs. According to Liu et al. (2015), the ABC classification system is suitable for classifying inventory according to several criteria, including its value to the company. Normally several demand items were held by the company with only a small amount being of such high value as to require close inventory control. In this case, generally, 5–15% of all inventory contributes to 70–80% of the total value (class A), while 15–30% contributes to 15–20% of total value (class B) and 50–60% of inventory contributes to 50–10% of total value (class C).

Such classification of items at the tube and pipe manufacturing company is accomplished by initially listing all items, their individual costs and estimated consumption. If a particular product had a stable demand, then historical consumption was used. The total value of the item was derived from multiplying its unit cost by demand and then rearranging the list in the order of the total value, from the highest value downwards.

The total value in sequence was then added to obtain the cumulative value. Using 10–20% of the total number starting with the highest at the top, the cumulative total value covered by these parts was found, which was from 70–80%, and according to adjusting the number of these items to obtain class A. An additional 15–20% of these parts was established to cover approximately 15–20% of the cumulative total value, over the total value for class A. In addition, the cumulative value for class B was established to be from 70–90%. Class C parts contributed the balance and these were approximately 10% of the extended value. The list of critical parts for production was revised and upgraded to class A or B. The following conclusions were drawn from the analysis:

- Class A: 26% of the total number of parts in stock, accounting for 49% of the total cost of the items in stock.
- Class B: 29% of the total parts in stock, accounting for 29% of the total value of the inventory in stock.
- Class C: 45% of the total quantities of inventory, accounting for 22% of the total value of items in stock.

This analysis showed that raw materials and components with work in progress as well as finished goods were not mixed since they should be controlled with a separate ABC classification. The derivation of key performance indicators (KPI) necessary for the design of the inventory management and control system, processes and functional areas was performed first to avoid data overload as a result of considering all the data as a KPI. Skilled operators for the production machinery were also engaged to ensure that the correct data were collated but ensuring that the processes did not influence or change their usual performance attributes. This was done by first briefing all who were involved to explain the purpose and that this was meant to improve the performance of the company.

The moving average for time series forecasting was used, taking the moving average MA(n) for the forecast period into the future using Equation (13.17), where n was the number of observations made in the forecast calculation and D was the actual demand observed for the historical period up to the time period t. Data and information for a two-year period were collated from the material demands and recorded for each product and plotted to establish the demand behaviour of which it was observed that bolts and nuts had the largest volume of quantities movement:

$$MA(n) = \frac{(D_t + D_{t-1} + \ldots + D_{t-n+1}}{n} \quad (13.17)$$

13.4.4 Design of the inventory control system

The design of the inventory management and control system was premised on the need to establish links between inventory decisions and the company's strategic goals in line with integrating the design office (graphic information) and production planning, scheduling and maintenance (non-graphic information). This required a good understanding of the company's markets, customer expectations as well as the inherent characteristics of the parts or spares in inventories throughout the value chain. The codification system was developed to apply a disciplined approach by which all parts of the service supply were identified and recorded in a uniform manner (Nyemba & Mbohwa, 2017). This was done in line with correctly identifying the locations of parts on the machine or in stock, based on the following principles:

- Establishment of a common language to ensure there is one unique stock number for each part of supply.
- Where the source of data was originating from, that is, whether technical information was required including a drawing, specification or standards from the manufacturer. (This was the key to establishing integration with the company's available CAD system.)

Based on the company's wide range of operations and historical data, albeit scanty, and the information gathered from the technical services and production departments, Table 13.1 shows the guidelines formulated and used in the developing codes for the inventory.

The control of inventory varied from one part to another depending on the importance of each part. Generally, there should be control mechanisms applied for the control of slow-moving and obsolete inventory. This was achieved by defining slow-moving, excess and obsolete materials by ensuring that there were reports that identified and measured these. The derivation of

Table 13.1 Guidelines used in developing the codification system

Rule/guideline	*Details*
1. Each part to have a unique number for identity (ID No.)	Important in ensuring that each part was uniquely identified and that there was no duplication even from same manufacturer.
2. ID No. to follow progression of the product flow	Made it easier to generate the numbers and easy to understand, for example, initial parts represented by lower numbers etc.
3. ID No. to follow the progression from bottom going up	Helps in identifying parts of the same type located in the same place
4. ID No. to follow progression from right to left, taking position where the operator stands while facing the machine	Identifies parts of the same type located in same place when difficult to give a number if Rule 4 fails to give ID No.
5. Description of parts should be as per manufacturer	To avoid confusion on which part is being referred to
6. Codification of all hand tools started with two zeros, for example, 0001208805	An arbitrary value that distinguished hand tools from machinery
7. Each part to maintain its Stores Number (S. No.)	To ensure compatibility of the new system with the old S. Nos.

(*Source:* Nyemba & Mbohwa, 2017)

Table 13.2 Description of levels for the 10-digit codification system

Level	*Description*
1	Described the machine on which the item was located, where the numbering of machines follows the production flow line
2	When considering various sections on the machine, the production flow from one section to another was taken
3	All machine assemblies and different mechanisms present on a machine, for example, all electrical equipment and pneumatic systems before breakup, started to be codified at this stage
4	Itemization of parts started at this stage, that is, shafts, sprockets, pinion etc.
5	Represented the final part of the sub-assembly which included bolts, washers, nuts, brushes etc.

(*Source:* Nyemba & Mbohwa, 2017)

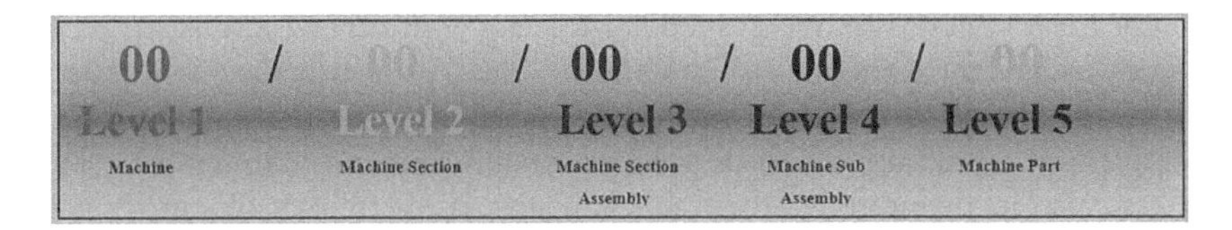

Figure 13.8 The 10-digit codification system levels.

(*Source:* Nyemba & Mbohwa, 2017)

the inventory codes for different parts within the production plant, their usage and flow were done by using the set of guidelines in Table 13.1 to obtain the 10-digit inventory codification for the inventory management and control system in conjunction with the generation of bills of materials and thus linking to the CAD system within the company. In addition, five levels were also formulated where every two consecutive digits represented one level as shown in Table 13.2. Figure 13.8 shows a schematic of the 10-digit codification system that was developed in this case study.

13.4.5 Results and implications

The development of the 10-digit inventory codification system for the tube and pipe manufacturing company was done in such a way as to simplify the identification of parts for use in an integrated system for linking graphic information generated in the technical and services department with non-graphic information used in production planning, scheduling and maintenance.

The implementation of this system at the tube and pipe manufacturing company led to effective planning for spare parts required for production machinery, coupled with ensuring that there were no duplicate spares that may already be in stock, ultimately ensuring that only required and appropriate spares were procured and issued. The five levels in the 10-digit inventory codification system were designed to identify Machine, Machine Section, Machine Section Assembly, Machine Sub-Assembly and Machine Part or Spare.

Microsoft Access was used in the development of the inventory database because of its compatibility and accessibility to the CAD system that the company employed at the time, that is, AutoCAD. Essentially, the codification system identified machine spare parts scheduling plan records, coupled with all pieces of equipment that a part can be used for at the inventory record level. The same system was used to prompt the Procurement and Stores department to automatically reorder stocks when necessary, based on the production schedule and not just to procure stocks for storage as was the case previously. The system also provided classes for easily and quickly identifying and locating spares coupled with the printing of bills of materials or catalogues for management reports.

A close analysis of the behaviour of all spares on the machines was also done through the reports generated by the database. The 10-digit inventory codification system was recommended for use with other established techniques such as the Kanban (pull) and MRP (push) as well as the CAD system at the company to form a fully integrated and hybrid system because of their complementary nature for optimized manufacturing and maintenance. The system also incorporated control mechanisms to ensure that inventory was appropriately managed in so far as decision making such as whether to buy or produce a spare part in-house.

On implementation of the system, the company was also advised of the need to rate suppliers based on their quality of service and delivery, adoption of engineering change management for staff in line with rapid changes in technology as well as hybrid shop floor control techniques that included a complete integration and computerization of the company. Replenishments of spare parts should be recorded in the database for individual machines to enable quantification of the average requirements for spares, also taking into cognizance, the importance and usage of standard spare parts, thus reliability in fitting and use as well as procurement.

The recommended integration and full computerization of all departments at the company would ensure that the procurement and stores department was in control of inventories in conjunction with the other departments that made use of the spares, unlike the situation prior to this research, where the procurement department worked as an entirely different entity to the other departments. The integration of such systems, including the CAD in technical services could also be done as recommended in previous chapters where, even external suppliers and customers are linked to improving throughput and competitiveness, with added network security of course.

The demand behaviour for most of the inventory parts in stock at the company was established to be seasonal although the variation periods differed from one spare to another. It was also established that in excess of 51% of parts in stock were unproductive as they were either slow-moving or obsolete. Their removal from stock reduced stock holding costs, apart from realizing some revenue from their disposal. Guidelines which were formulated to develop the 10-digit codification system enabled users to allocate codes to any new spare parts and these were coded in conjunction with the company's old numbering system for continuity as well as to ensure that no part was omitted.

A cost-benefit was also carried out and revealed that, although the implementation of such a new system would be costly, recommendations made in line with the 10-digit system resulted in reductions of inventory through removal and disposal of slow-moving and obsolete stock. The implementation of the system also resulted in not only the improvement of record-keeping and accountability but also a decrease in maintenance and machine downtimes. The system provided simplification in maintenance enhancement of quality and reduction in lead times, hence increased profits and viability. The added advantage of the 10-digit codification system was the integration with other systems at the company, including compatibility and ability to be linked with the AutoCAD system at the company.

13.5 SUMMARY

Apart from the three case studies in previous chapters, this chapter focused on three further case studies to demonstrate the importance and use of CAD systems in engineering design and manufacture. The first case study demonstrated how AutoLISP was used to model and simulate a four-bar chain mechanism commonly used in windscreen wipers, to enable analysis and optimization of engineering designs. The second case study used the example of a dump truck tipping mechanism but this time using EdenLISP, a derivative of AutoLISP, which combines aspects of both procedural and functional programming languages' attributes. The third case study demonstrated the integration of CAD systems with other functional departments within a company through the implementation of a 10-digit inventory codification system for inventory management and control and also facilitating the link between graphic information (drawings) and non-graphic information such as production planning and scheduling for maintenance.

13.6 REVIEW EXERCISES

1. AutoLISP allows users to develop macro-programs for graphical applications. Outline how a typical mechanism can be broken down, modelled and simulated in AutoCAD.
2. EdenLISP was developed to combine both aspects of procedural and functional programming languages. Explain how these two attributes

simplify programming in LISP by focusing on the major differences between AutoLISP and EdenLISP.

3. The 10-digit codification system was designed to control inventory and assist management in decision-making as well as production scheduling and maintenance. Outline the importance of integration of company systems and the linking of graphic and non-graphic information in a typical engineering and manufacturing company.

Chapter 14

Typical examination questions

QUESTION 1

a) Describe the minimum hardware and software requirements for a modern CAD system in terms of input, processing and output.
[5 marks]

b) There are several ways to input information in a typical CAD system such as AutoCAD. Explain with the aid of sketches the three most commonly used approaches.
[5 marks]

c) While it is quicker for engineering designers to use icons and slide menus in CAD, explain why it is vital for the same personnel to have a grasp of all the techniques employed in inputting information, particularly using the Command Prompt.
[5 marks]

d) List and explain seven advantages and three limitations (disadvantages) of using CAD systems compared to manual drawing.
[10 marks]

e) Figure 14.1 shows a pictorial (isometric) view of a cast iron bracket.

i. List the minimum number of primitives required to produce such a model in a typical CAD system such as AutoCAD.
[5 marks]

ii. Describe in stages and using sketches, how such a model can be converted into three orthographic projections in a CAD system.
[5 marks]

DOI: 10.1201/9781003288626-14

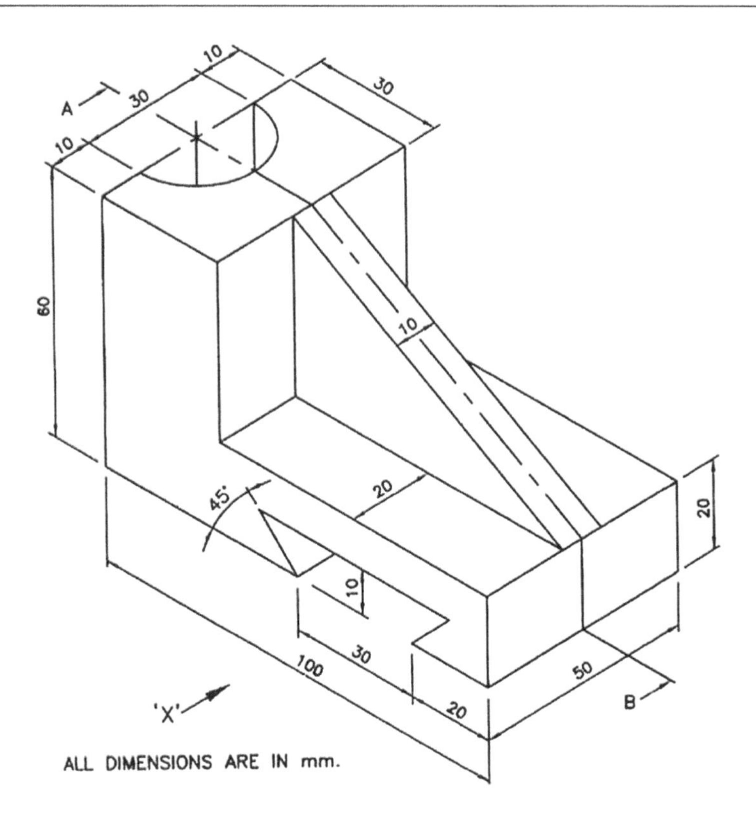

Figure 14.1 Cast iron bracket.

(iii) In third angle projection, produce the Front view taken in the direction X, Right side view and the Plan as they would appear on a typical CAD system such as AutoCAD.

[Total: 40 marks]

QUESTION 2

a) As a newly appointed Product Development Engineer at a design and manufacturing company that utilizes conventional machine tools and traditional methods of design using the drawing board, write a brief proposal to advise to convince management to migrate to Computer Aided Design and Manufacture. Focus your answer on 10 reasons to sufficiently convince the company to invest in the new technology.

[10 marks]

b) Coupled with your proposal in (a), include a brief precaution to the company on some of the challenges that can be encountered so that the company is well informed and better prepared for any eventuality if they encounter such challenges.

[5 marks]

c) CAD Systems such as AutoCAD can be customized to suit the applications in any discipline of Engineering. Using your discipline, list and explain five attributes or specific areas that can benefit from such CAD systems.

[5 marks]

d) Figure 14.2 shows a pictorial drawing of a portion of a pumping station. Extracting and developing the technical drawings and details for such a plant manually can be quite complex and in some cases, errors can be generated, even when the maximum size of paper such as A0 is used.

i. Describe briefly and explain some of the challenges encountered in the manual development of technical drawings.

[10 marks]

ii. Study the figure carefully and then explain how a CAD system such as AutoCAD can overcome some of these challenges. Support your answer by identifying some of the components of the pumping station and use these to explain your answer.

[10 marks]

[Total 40 marks]

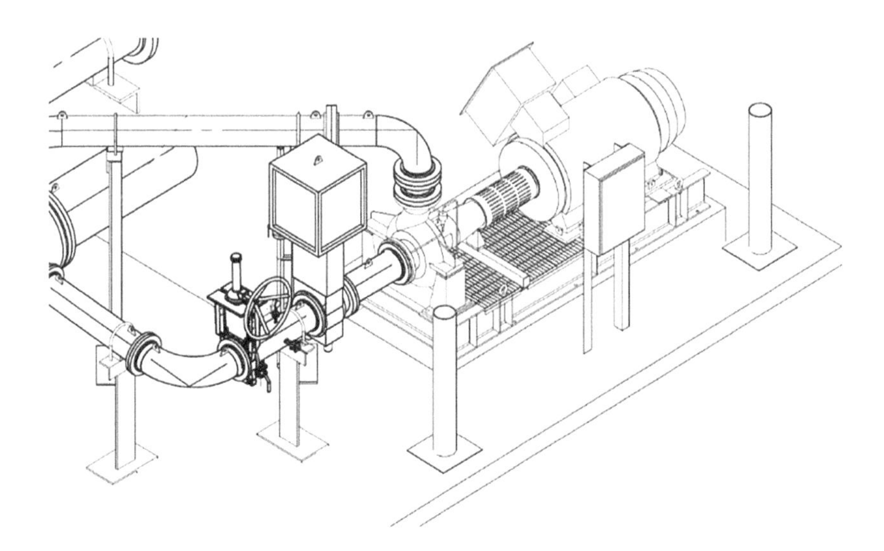

Figure 14.2 Pictorial view of a portion of a pumping station.

QUESTION 3

a) Mechanical Engineering designs mainly consist of machine drawings along with various components (elements). Identify five such components that can be drawn in a CAD system, stored in a library and retrieved when required in a drawing. For each of the five blocks chosen, specify the important aspects that should be defined along with the blocks.

[10 marks]

b) There are two general ways in which blocks can be created in a CAD system such as AutoCAD. What are the fundamental differences between a WBLOCK and a BLOCK as employed in AutoCAD? Use a typical example where these two would be most useful.

[5 marks]

c) The use of blocks as symbols in CAD libraries is meant to avoid the repetitive reproduction of standard components used in engineering. Using the following typical components in Figure 14.3 that are commonly used as blocks, describe the important attributes that should be attached to these symbols for easy retrieval from a catalogue of symbols in a CAD database for use in a drawing.

[10 marks]

c) Explain what the AutoCAD system variables ATTREQ and ATTDIA are used for in the process of creating, storage and retrieval of blocks from a CAD database. The two system variables are usually set in binary notation (0 or 1). Explain what this means for each of them being set at 0 or 1.

[5 marks]

d) Figure 14.4 shows an exploded view of sports bike frame along with various components for its assembly. Select 10 of the components shown and using the numbers provided, create a table and insert the following details to fill Table 14.1.

[10 marks]

[Total 40 marks]

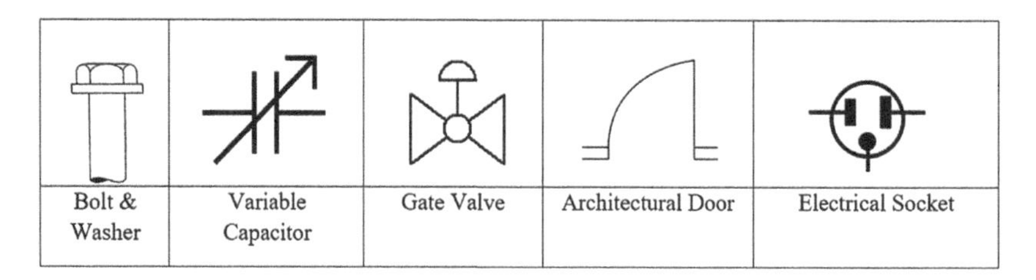

Figure 14.3 Commonly used symbols (blocks) in engineering designs.

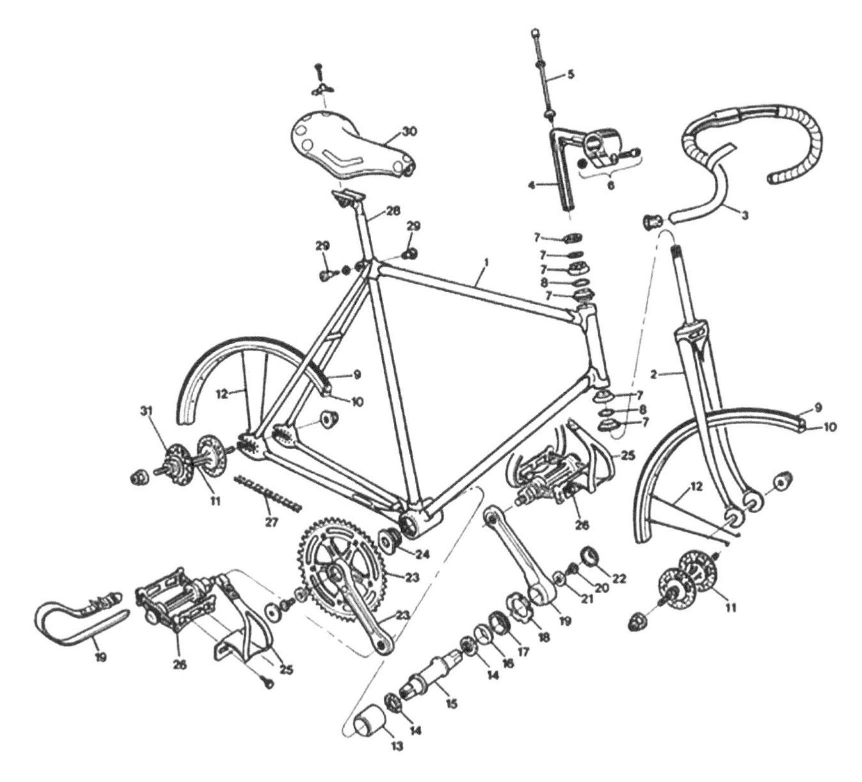

Figure 14.4 Exploded view of a sports bike frame assembly.

Table 14.1 Components of a sports bike to be created and stored as blocks

Part number	*Description/name*	*Insertion point*	*Attributes to attach*

QUESTION 4

a) Engineering drawings and designs are often presented in different formats such as orthographic projections, isometric projections and three-dimensional models. Explain the differences among these formats and for each provide a practical everyday scenario where each one would be more applicable.

[5 marks]

b) Three-dimensional models can be generated in a typical CAD system using different methods depending on what needs to be modelled.

Explain the difference between boundary representation and (B-Rep) and constructive solid geometry (CSG) as applied in the development of 3D models, giving one practical example for each of the two techniques.

[10 marks]

c) Figure 14.5 shows model views of typical engineering structures/ objects in the form of a stepped shaft and walls of a building. Describe the most optimal way in which these 3D models can be produced using a CAD system such as AutoCAD.

[10 marks]

d) The major advantage of migrating from manual drawings to CAD lies in the inherent ability to modify drawings under development. Explain the use of the following modification commands in a typical CAD system using illustrated examples, paying particular attention to the correct procedure in their use:

(i) Explode
(ii) Rotate
(iii) Scale
(iv) Extend
(v) Trim

[5 marks]

e) Using schematics, describe the differences and applications of the three techniques employed in 3D modelling, that is, wireframe, surface and solid modelling.

[5 marks]

f) Several techniques are employed in 3D surface modelling. Describe each of the following techniques and for each include an example where it is most applicable.

(i) 3DFACE
(ii) EDGESURF
(iii) RULESURF

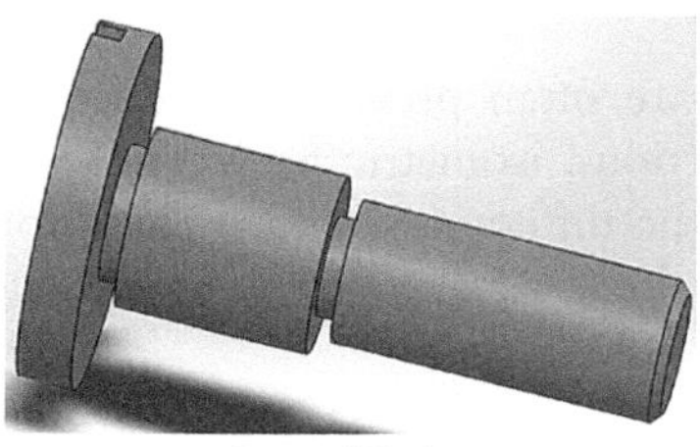

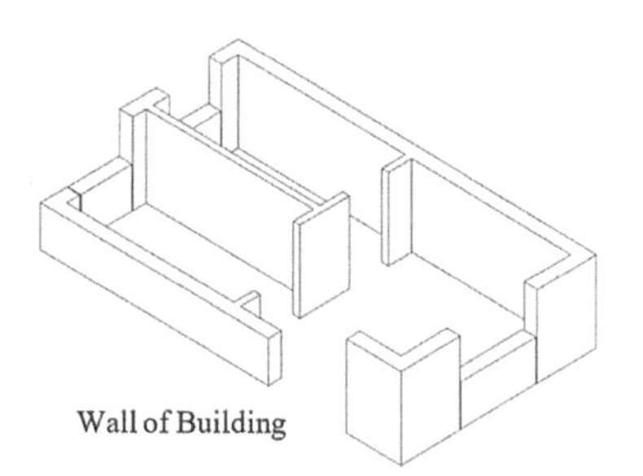

Figure 14.5 Typical models of engineering structures/objects.

(iv) TABSURF
(v) REVSURF

[5 marks]

[Total 40 marks]

QUESTION 5

a) List and explain the seven types of dimensions used in a typical CAD system such as AutoCAD.

[5 marks]

b) Using a sketch, describe five parameters required to properly and adequately insert a horizontal dimension in AutoCAD.

[5 marks]

c) Explain why it is often necessary to produce drawings with tolerances. Show this with the aid of sketches on how you can accomplish it in a CAD system.

[5 marks]

d) Explain using sketches, three types of arrows used in dimensioning and for each, describe the discipline where they are most applicable.

[5 marks]

d) Figure 14.6 shows a schematic of an orthographic projection. Using only one of the seven dimension types, in line with your answer to Question 5a), insert these into the schematic to completely describe the orthographic view in a CAD system.

[10 marks]

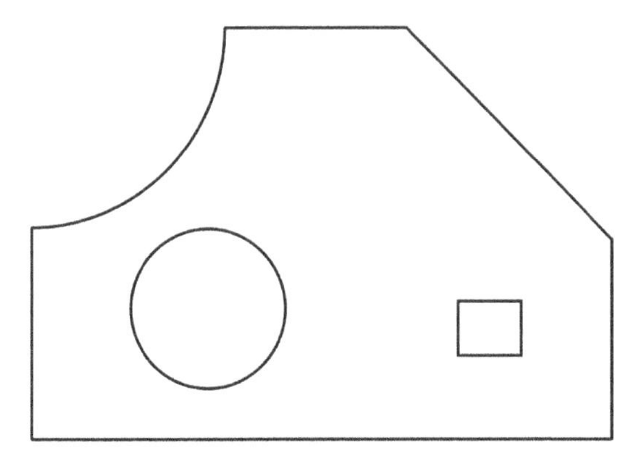

Figure 14.6 Orthographic view of an object.

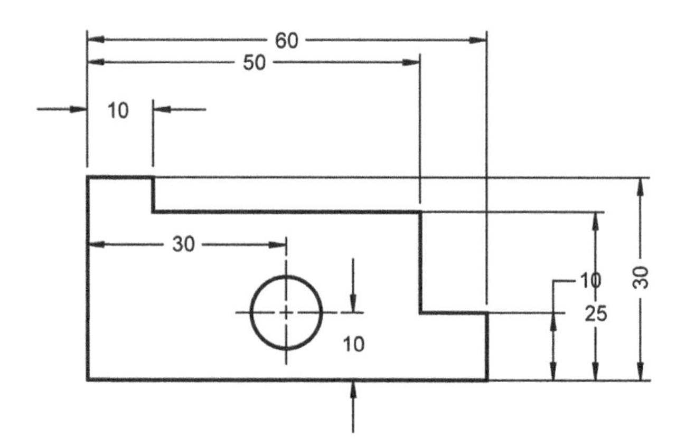

Figure 14.7 Incorrectly dimensioned drawing.

e) Figure 14.7 shows an orthographic view of an object that is incorrectly dimensioned in CAD, according to standard engineering drawing principles. Identify these mistakes and for each, provide the correction of what parameters need to be adjusted and then correctly dimension the view using AutoCAD.

[10 marks]

[Total 40 marks]

QUESTION 6

a) Using AutoCAD, create the seven architectural symbols shown in Figure 14.8 using the dimensions provided and save them using the names provided and store them in a suitable folder (catalogue) on your computer.

[10 marks]

b) Create the following layers: Walls (Red), Symbols (Yellow) and Dimensions (Blue), then produce the outline of an architectural plan as shown in Figure 14.9 and save it to a suitable AutoCAD directory folder.

[15 marks]

c) Retrieve and insert the symbols from the catalogue created in (a) to produce the completed architectural plan as shown in Figure 14.10, using the three defined layers in (b). Include all necessary but no superfluous dimensions.

[15 marks]

[Total 40 marks]

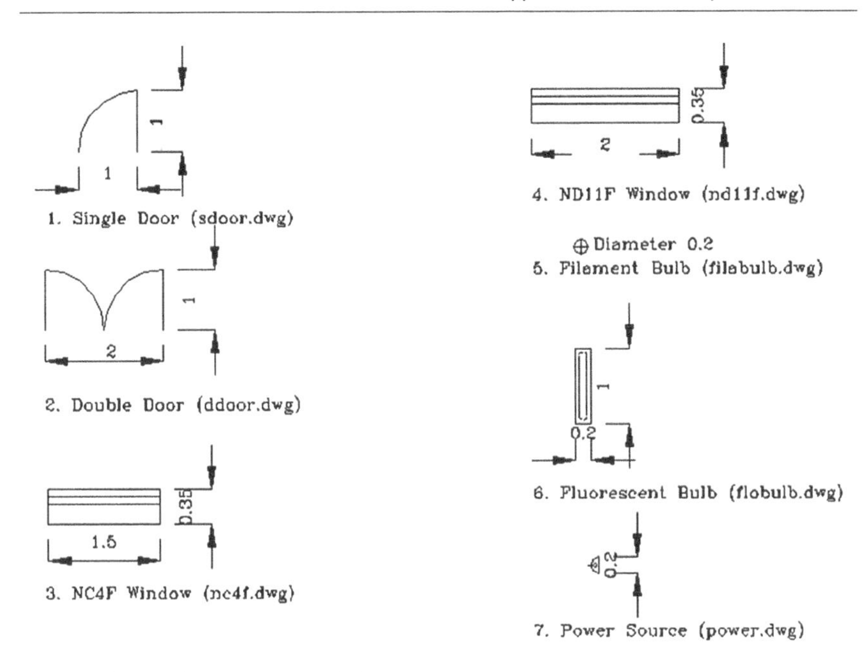

Figure 14.8 Architectural symbols.

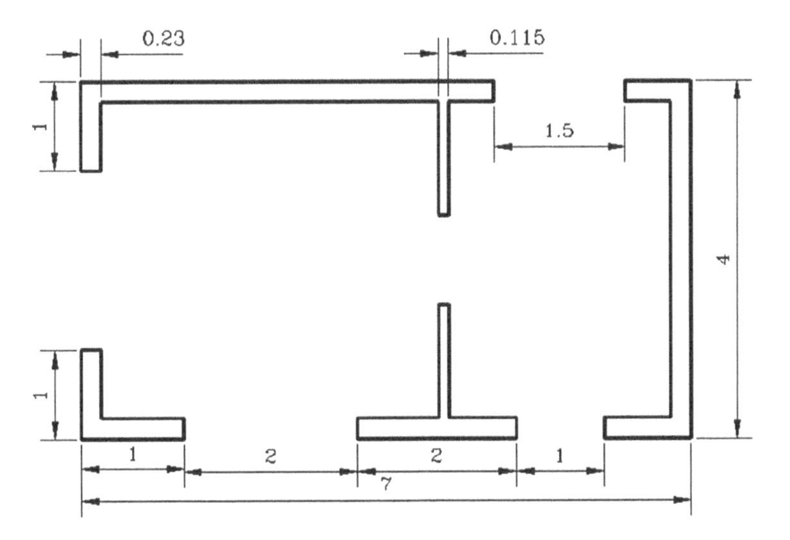

Figure 14.9 Dimensioned architectural walls of a building.

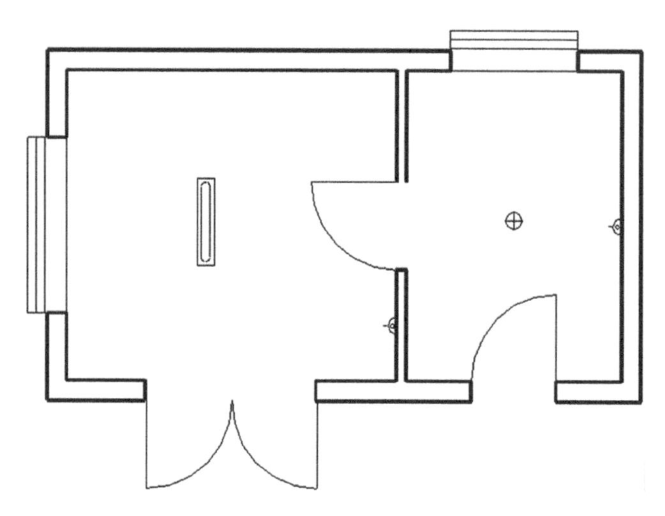

Figure 14.10 Architectural plan of a building.

QUESTION 7

a) Figure 14.11 shows a typical detailed engineering drawing of the plant layout and production flow processes for a motor vehicle manufacturing and assembling plant created using AutoCAD, albeit complex and difficult to read.

(i) Explain how this drawing can be simplified using CAD and in particular, the importance of the use of layers in drawings under development in CAD.

[10 marks]

(ii) Identify parts of this drawing that can be split into layers and explain how these can be distinguished among them.

[10 marks]

[Total 20 marks]

QUESTION 8

a) Distinguish between oblique and isometric projections as utilized in CAD systems.

[5 marks]

b) The most convenient method of drawing isometric views of any object in CAD is the boxing method. Describe the four steps by which

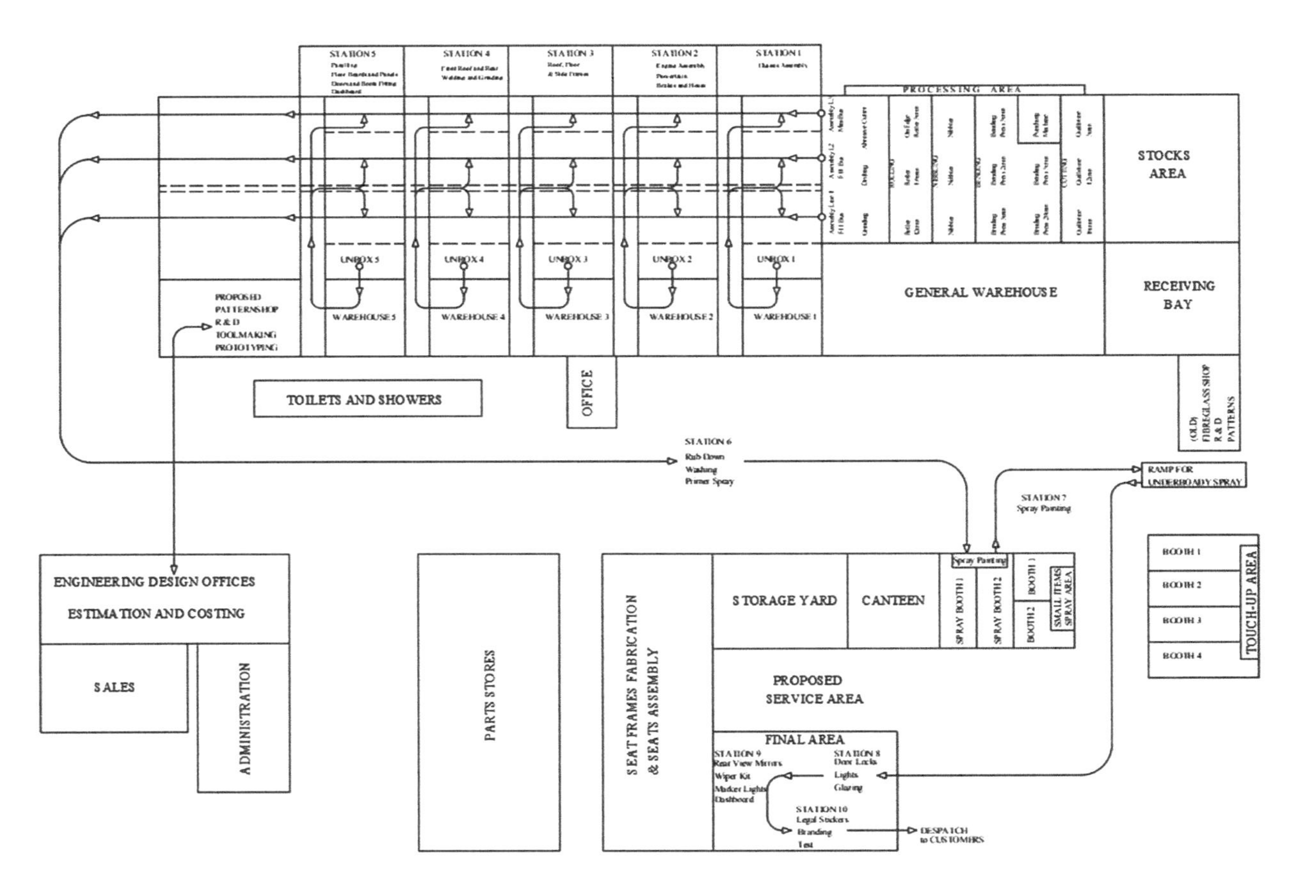

Figure 14.11 Plant layout and production flow of a motor vehicle manufacturing plant.

orthographic views can be transformed into isometric views using the boxing method.

[5 marks]

c) Although isometric views appear as three-dimensional objects, they are in fact two-dimensional objects. Explain why this is so, by clearly distinguishing between 2D and 3D objects.

[5 marks]

d) Figure 14.12 shows orthographic views of an engineering casting in first angle projection. Using the boxing method convert these to an isometric view using AutoCAD.

[5 marks]

[Total 20 marks]

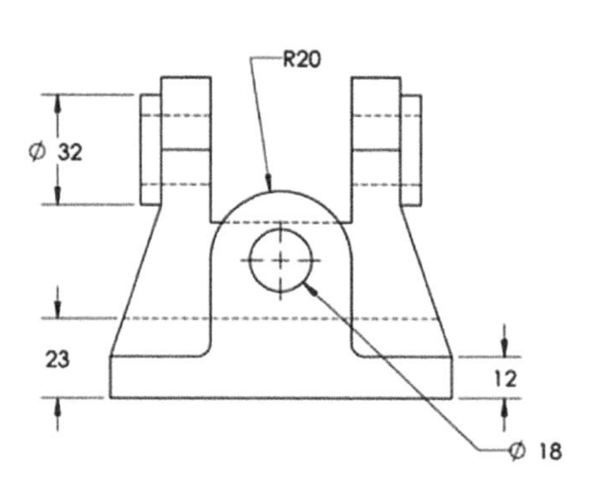

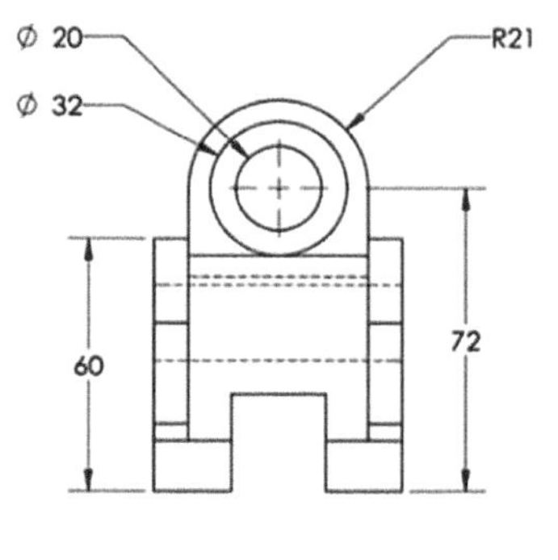

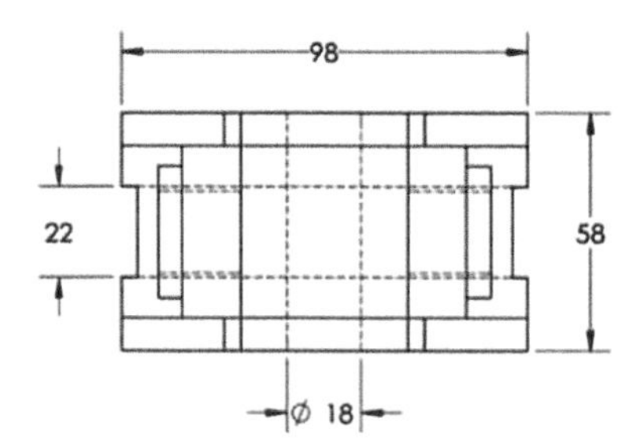

Figure 14.12 Orthographic views of an engineering casting.

QUESTION 9

a) List the six most commonly used primitives in 3D constructive solid geometry (CSG) modelling and for each state the parameters required for each to be drawn.

[5 marks]

b) Explain how 3D solid primitives are combined using Boolean algebra to obtain 3D solid models.

[5 marks]

c) Figure 14.13 shows a 3D model of a cast iron bracket. Study the figure carefully, then answer the following questions:

(i) What is the optimum number of solid primitives required to produce the 3D solid model of the cast iron bracket using CSG?

[5 marks]

(ii) Using clear illustrations or AutoCAD, demonstrate how the primitives in (i) are combined using Boolean algebra to obtain the 3D model as shown.

[5 marks]

[Total 20 marks]

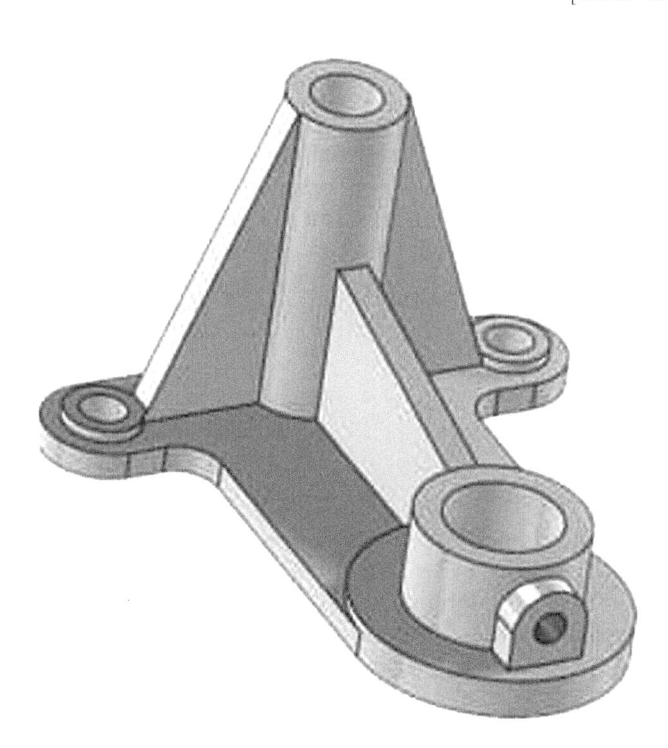

Figure 14.13 3D model of a cast iron bracket.

QUESTION 10

(a) The management of engineering drawings from CAD systems is critical as much as it is important to develop them. Developed drawings can be availed in different formats for flexibility. Explain how the following exchange formats are used in AutoCAD:

(i) WMF
(ii) PLT
(iii) DWF
(iv) EPS
(v) PDF

[5 marks]

(b) Using illustrations, distinguish and explain the differences between Model Space and Paper Space, focusing on the differences in the use of viewports in managing outputs or plotting drawings from each space environment.

[5 marks]

(c) List and explain five of the most important settings that should be configured before a CAD drawing can be plotted/printed on a regular size of paper.

[5 marks]

(d) Figure 14.14 shows a completed drawing in Paper Space ready for printing. However, this is not configured optimally to make the best use of resources. Explain how you reconfigure the system for an optimal layout on an appropriate size of paper.

[5 marks]

[Total 20 marks]

QUESTION 11

a) The major advantage of migrating from manual engineering drawings to Computer Aided Design (CAD), is the ability to reuse drawings such as standard symbols.
   (i) Describe how CAD systems store blocks for use in later tasks and how users can easily manage the blocks within the storage libraries.

[5 marks]

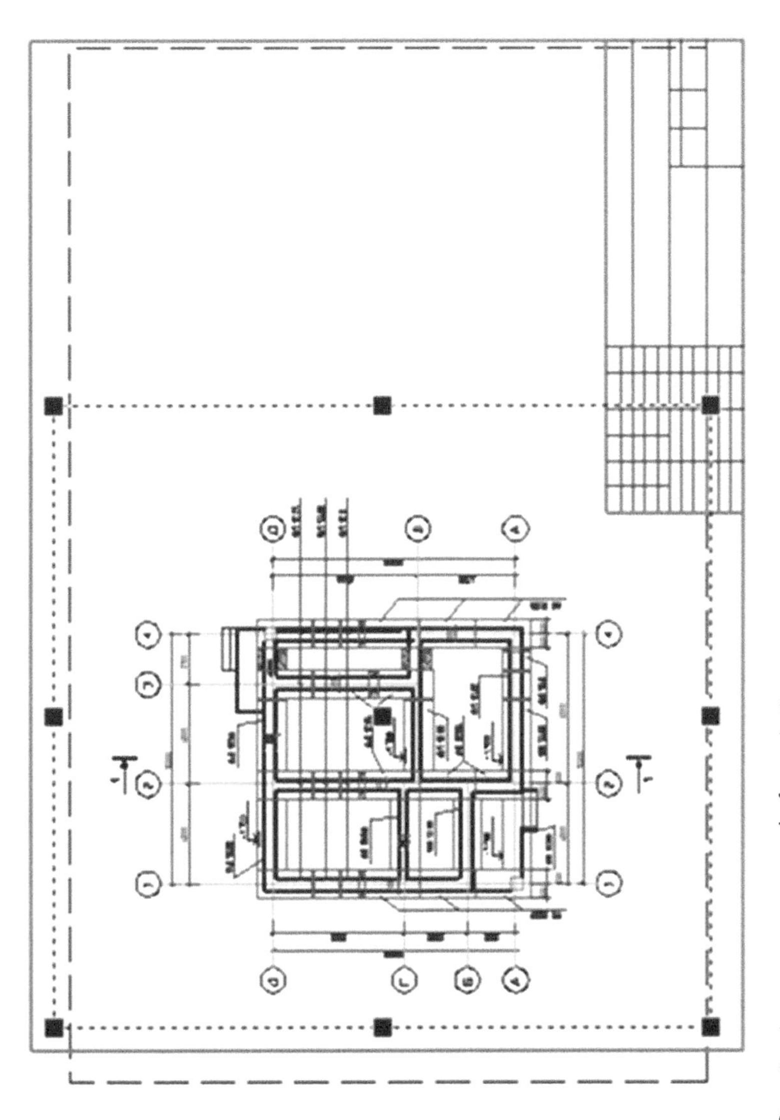

Figure 14.14 Drawing in paper space ready for printing.

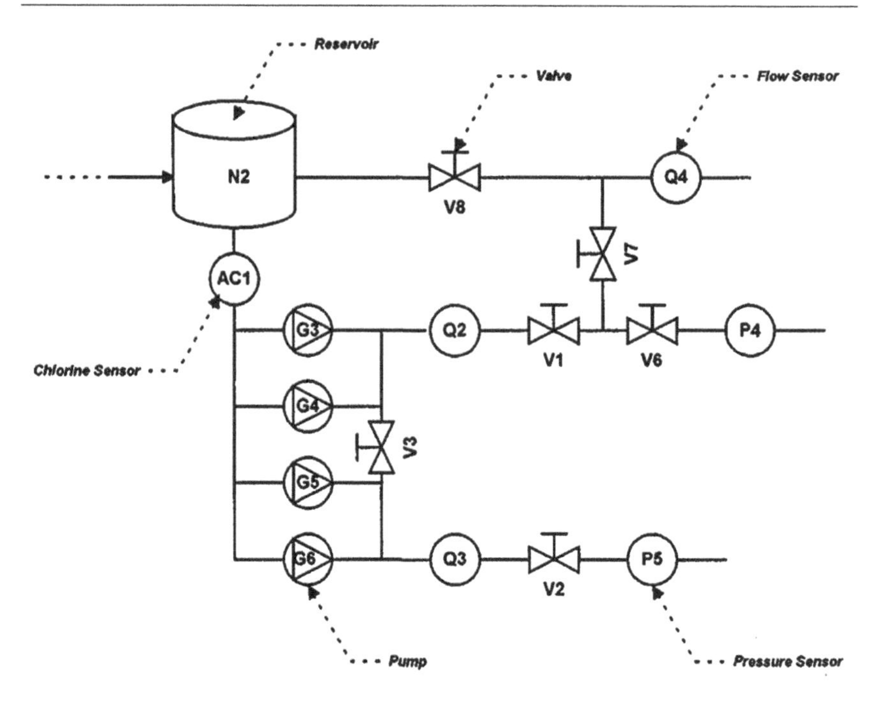

Figure 14.15 Model layout of a pumping station.

(ii) Explain the retrieval process for such blocks and their insertion in drawings.

[5 marks]

b) Figure 14.15 shows a model layout of a pumping station showing various components such as valves, pumps etc. As the plant engineer for the pump station:

(i) Explain the procedures that are followed by a CAD system to convert graphic information to non-graphic information such as the Bill of Materials (BoM).

[5 marks]

(ii) Draw up a typical BoM for the pumping station and insert suitable attributes.

[5 marks]

[Total 20 marks]

QUESTION 12

a) Modelling objects in engineering is to enable the analysis of products under development. Explain with the aid of illustrations, at least five such analyses that can be carried out using CAD to avoid prototyping of products under development.

[5 marks]

b) Explain the differences among the following visual styles in AutoCAD:

(i) 3D Wireframe
(ii) 3D Hidden
(iii) Realistic
(iv) Conceptual

[5 marks]

c) Figure 14.16 shows the 3D conceptual model of a coupling along with its orthographic projections, automatically generated and placed in third angle projection from the 3D model and displayed in four viewports of a CAD system.

(i) Outline the step-by-step procedure on how the orthographic projections were generated from the 3D model in a typical CAD package such as AutoCAD.

[5 marks]

(ii) Draw up a process plan to outline how the 3D model is initially generated in Model Space and then eventually into Paper Space as shown for printing.

[5marks]

[Total 20 marks]

QUESTION 13

a) Circles in isometric planes appear as ellipses. Explain how this can be achieved from one isometric plane to another.

[5 marks]

b) Using AutoCAD, develop a drawing template and title block for an A3 paper size in landscape format, then load it to proceed to (c).

[5 marks]

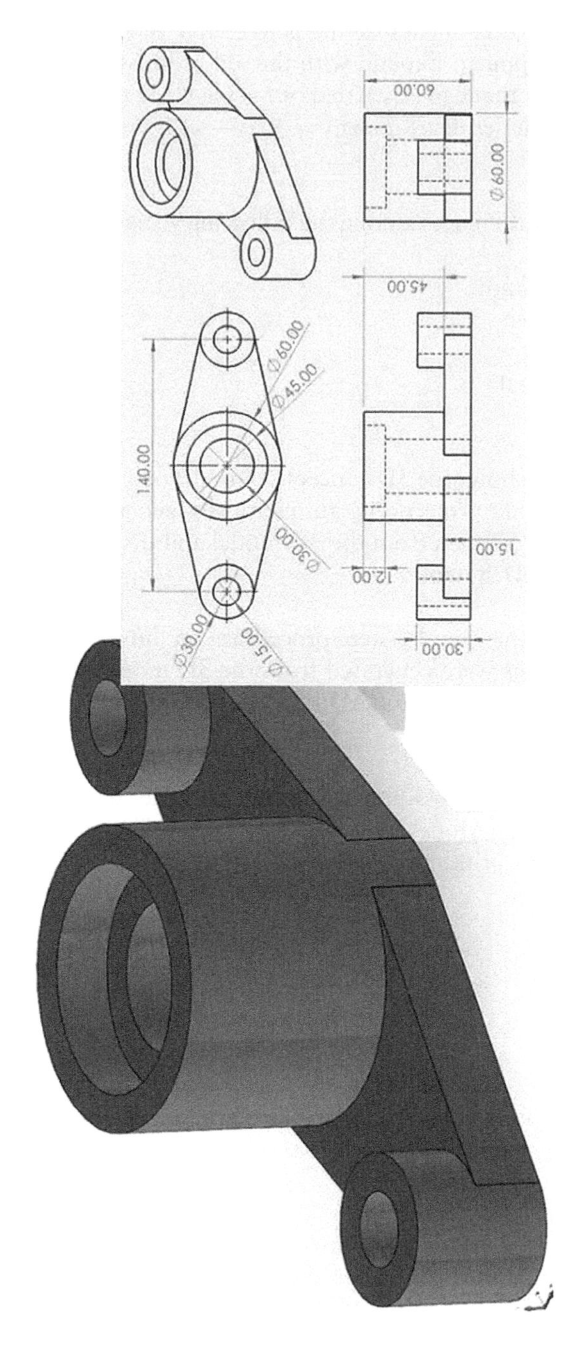

Figure 14.16 3D model of coupling with orthographic views.

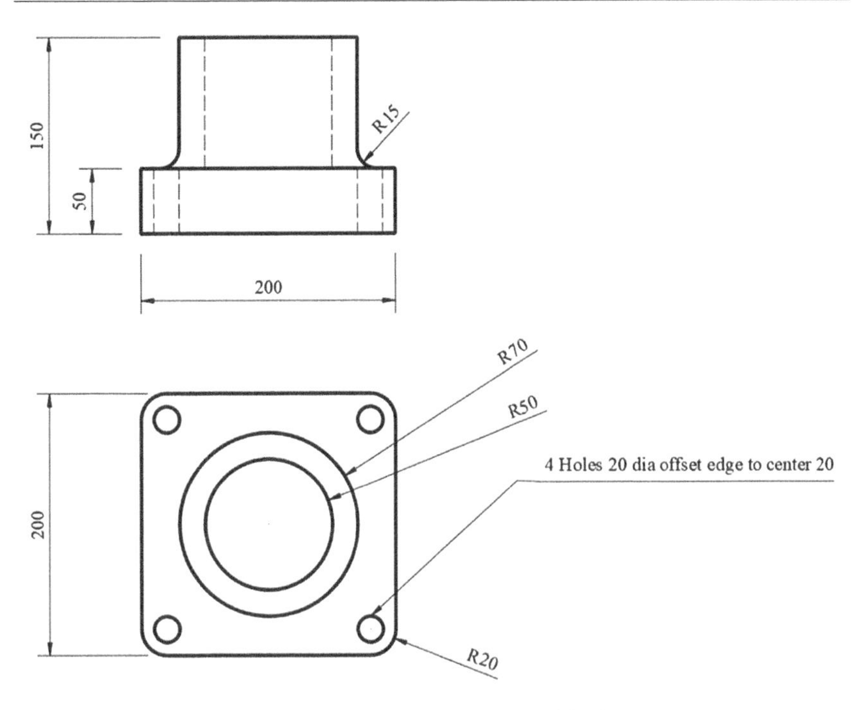

Figure 14.17 Orthographic views of an engineering block.

c) Using the orthographic views shown in Figure 14.17 and within the title block from (b), generate an isometric box that will accommodate the views provided, including the Left Side View visible as well.

[5 marks]

d) Complete the isometric view using the boxing method and the two views shown in Figure 14.17, clearly showing the isometric planes, Top, Front and Left Side.

QUESTION 14

a) 3D surfaces can be generated on objects that are bound by three or four edges. Explain how the PFACE command in AutoCAD is used to generate surfaces on surfaces that are bound by five or more edges.

[5 marks]

b) Figure 14.18 (left) shows a pictorial view of an engineering block. Using AutoCAD, construct the wireframe model of the block.

[10 marks]

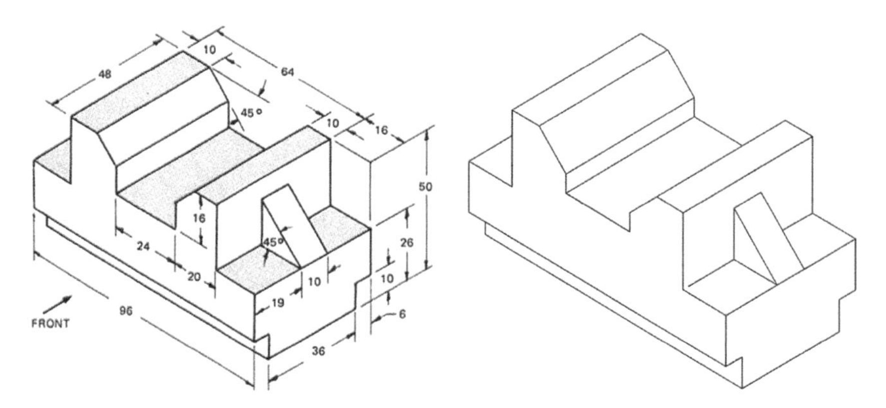

Figure 14.18 3D surface model of an engineering block.

c) Using the wireframe model in (b), cover the model with surfaces using the 3DFACE or PFACE commands, then apply the HIDE command to obtain the model shown in Figure 14.18 (right).

[5 marks]

[Total 20 marks]

Appendix A1

Selected program listings for the bom utility module for Chapter 7

APPENDIX A1.1

```
///BOM_MENU.MNU
/// Bill of Materials partial menu file.

***MENUGROUP=bom_menu
***POP1
ID_Title [/BBOM]
ID_Entry [/SSingle Entry...]^C^Cbom_entry
ID_Compile [/CCompile BOM...]^C^Ccompile_bom
ID_Help [/HHelp on BOM...]^C^C(help "bom_help.ahp"
"general_help")
ID_About [/AAbout the BOM Utility...]^C^C(help "bom_help.ahp"
"about_bom")
[--]

***ACCELERATORS
ID_Help [SHIFT+"F1"]
ID_About [SHIFT+"A"]
```

APPENDIX A1.2

```
;;; BOM.LSP

(defun c:bom()
       (setvar "filedia" 0)
       (command "menuload" "bom_menu")
       (menucmd "P11=+bom_menu.pop1")
       (setvar "filedia" 1)
)
(princ "\nBOM menu loaded.\n")
(princ)
```

APPENDIX A1.3

```
;;;  COMPILE_BOM.LSP
;;;  AutoLISP file for the "Compile BOM" menu

(defun do_compile ()
  (setq outfile (getfiled "Enter the Output File" "" "txt" 9))
  (setvar "filedia" 0)
; disabling appearance of file dialog box
  (command "attext" "C" "bom_tmp1" outfile)
  (setvar "filedia" 1)
  (princ (strcat "\nBill of Materials successfully exported
   to " outfile))
)

;  BOM ERROR HANDLER
;  Defines the error handler for Compilation
(defun bom_error (msg)
  (setq *error* olderr)
  (if (not comperr)
    (princ (strcat "\nBOM compilation error: " msg))
    (princ bomerr)
  )
  (princ)
)
; main routine

(defun C:COMPILE_BOM ()
  (setq olderr *error*
      *error* bom_error
      comperr nil
  )
  (setq sblip (getvar "blipmode"))
  (setq scmde (getvar "cmdecho"))
  (setvar "blipmode" 0)
  (setvar "cmdecho" 0)
  (do_compile)
  (setvar "blipmode" sblip)
  (setvar "cmdecho" scmde)
  (setq *error* olderr)
  (princ)
)

(princ "\nCompile_bom successfully loaded")
(princ)
```

APPENDIX A1.4

```
;;; ACAD.LSP
;;; Loaded automatically when AutoCAD starts.
```

```
;;; Loads "COMPILE_BOM,LSP" ,"BOM.LSP" and ;"BOM_ENTRY.LSP"

(autoload "bom" '("bom"))
(autoload "bom_entry" '("bom_entry"))
(autoload "compile_bom" '("compile_bom"))
(princ)
```

APPENDIX A1.5

```
/// ATT_TXT_PROP.DCL
/// DCL file for BOM_ENTRY.LSP. Defines ///the dialog box
for entering attribute text
/// properties
///
txt_prop : dialog {
  label = "Attribute Properties";
  : boxed_column {
    label = " ";
    : edit_box {
      label = "&Text Height";
      key = "dl_txt_ht";
      edit_width = 4;
    }
    : edit_box {
      label = "&Rotation Angle (deg)";
      key = "dl_rot_ang";
      edit_width = 4;
    }
  }
  : row {
    : spacer { width = 1; }
    : button {
      label = "OK";
      key = "accept";
      width = 8;
      fixed_width = true;
    }
    : button {
      label = "Cancel";
      is_cancel = true;
      key = "cancel";
      width = 8;
      fixed_width = true;
    }
    : button {
      label = "Help";
      key = "help";
      width = 8;
      fixed_width = true;
```

```
    }
    : spacer { width = 1;}
  }
}
```

APPENDIX A1.6

```
/// BLK_NAME.DCL
/// DCL file for BLK_NAME.LSP Defines the ///dialog box for
entering item block name
///
blk_box : dialog {
  label = "Item Block Name";
  : edit_box {
    label = "&Block Name (31 letters max., no spaces)";
    key = "dl_blk_name";
    edit_width = 31;
  }
  : row {
    : spacer { width = 1; }
    : button {
      label = "OK";
      key = "accept";
      width = 8;
      fixed_width = true;
    }
    : button {
      label = "Cancel";
      is_cancel = true;
      key = "cancel";
      width = 8;
      fixed_width = true;
    }
    : button {
      label = "Help";
      key = "help";
      width = 8;
      fixed_width = true;
    }
    : spacer { width = 1;}
  }
}
```

APPENDIX A1.7

```
/// ATTR_INFO.DCL
/// DCL file for BOM_ENTRY.LSP Defines the ///dialog box for
entering attribute information
```

```
///
att_box : dialog {
  label = "Item Identifying Attributes";
  : boxed_column {
    label = " ";
    : edit_box {
      label = "&Item number";
      key = "dl_item_no";
      edit_width = 5;
    }
    : edit_box {
      label = "&Description";
      key = "dl_descrip";
      edit_width = 35;
    }
    : edit_box {
      label = "&Material";
      key = "dl_matrl";
      edit_width = 10;
    }
  }
  : row {
    : spacer { width = 1; }
    : button {
      label = "OK";
      key = "accept";
      width = 8;
      fixed_width = true;
    }
    : button {
      label = "Cancel";
      is_cancel = true;
      key = "cancel";
      width = 8;
      fixed_width = true;
    }
    : button {
      label = "Help";
      key = "help";
      width = 8;
      fixed_width = true;
    }
    : spacer { width = 1;}
  }
}
```

APPENDIX A1.8

```
TEMPLATE OUTPUT FILE
Used in Defining Rows and Columns and the Associated
Attributes.
ITEM          N005000
BASIC_QTY     N005000
DESCRIP       C035000
MATERIAL      C010000
SECTION       C015000
LENGTH        N006002
STORES_CODE   C010000
PRICE         N010002
```

Appendix A2

Selected program listings for mechanisms for Chapter 13

APPENDIX A2.1

AutoLISP Program for Modelling and Simulating a 4-Bar Linkage Mechanism

```
(defun c:4bar (/ l1 l2 l3 l4 theta2 theta3 theta4 pt1 pt2
pt3 l beta psi lambd
     a b c sp1 sp2 ep1 ep2 tp ans stt hp1 hp2)
     (defun getinput()
        (setq pt1 (getpoint "enter origin of crank: ")
          l2 (getdist pt1 "enter length of the crank: ")
          theta2 (getreal "enter starting angle of crank in
          radians: ")
          pt4 (getpoint "enter position of the end of four
          bar linkage: ")
          l3 (getreal "enter length of the coupler: ")
          l4 (getreal "enter length of the rocker: ")
          l1 (distance pt1 pt4)
        )
      )
      (getinput)
      (cond ((and (<= l1 (+(- l3 l4) l2))(>= l1 (-(+ l3 l4)
      l2)))
          (prompt "error: ensure (L3-L4+L2<L1<L3+L4-L2),
please reenter: ")
          (getinput)
         )
         (T (prompt "Resulting Link- Rotation of crank and
         oscillation of rocker "))
      )
      (command "zoom" "w" (polar pt1 (/(* 5 pi) 4) (sqrt(* 2
      (* l2 l2))))
               (polar pt4 (/ pi 4) (sqrt(* 2 (* l4 l4))))
      )
```

```
; sp and ep define the start and end of an arc that will
locate the restraints.
; hp defines the window corner point to enable hatching of
the restraints.
; Also note that the angle defining the arc is entered in
degrees and not
; radians as in normal AutoLISP functions.
      (setq sp1 (polar pt1 pi (/ l2 10))
          ep1 (polar pt1 0.0 (/ l2 10))
          sp2 (polar pt4 pi (/ l2 10))
          ep2 (polar pt4 0.0 (/ l2 10))
          hp1 (polar pt1 (/ pi 4) (/ l2 5))
          hp2 (polar pt4 (/ pi 4) (/ l2 5))
      )
      (command "line" sp1 ep1 "")
      (command "arc" sp1 "e" ep1 "a" -180)
      (command "line" sp2 ep2 "")
      (command "arc" sp2 "e" ep2 "a" -180)
      (command "hatch" "grate" "" 45 "w" sp1 hp1 "")
      (command "hatch" "grate" "" 45 "w" sp2 hp2 "")
; tp defines the positioning of text on the screen.
      (setq tp (polar pt1 (/ pi 2) (sqrt(* 2 (* l4 l4)))))
; The font style is changed from the default standard to
multi-stroke.
      (command "style" "multi-stroke" "romanc" "" "" "" ""
      "" "")
      (command "text" tp (/ l2 10) "" "FOUR-BAR LINKAGE
      MECHANISM" )
; The remainder of this program is contained in a loop,
starting with the
; angle of crank at 0.0 radians, the three links are
displayed. The display
; is cleared and the angle of crank incremented by 0.1
radians.
     (while (>= theta2 0.0)
       (setq pt2 (polar pt1 theta2 l2)
          l (sqrt(-(+(* l1 l1)(* l2 l2))(* l1 l2)(cos
          theta2)))
          a (*(/ l2 l)(sin theta2))
       )
; To avoid the 'divide by zero' error the angles at which
this occurs are
; pre-defined.
       (setq beta (cond ((equal a 1.0 0.00001)(/ pi 2))
                  ((equal a -1.0 0.00001)(/(* 3 pi) 2))
                  (t (atan(/ a (sqrt(- 1 (* a a))))))
                )
        )
        (setq b (/(-(+(* l l)(* l3 l3))(* l4 l4))(* 2 l
        l3)))
```

```
        (setq psi (cond ((equal b 0.0 0.00001)(/ pi 2))
                        (t (atan (/(sqrt(- 1 (* b b))) b)))
                  )
        )
        (setq c (* (/ 13 14) (sin psi)))
        (setq lambd (cond ((equal c 1.0 0.00001)(/ pi 2))
                          ((equal c -1.0 0.00001)(/(* 3 pi)
                          2))
                          (t (atan (/ c (sqrt(- 1 (* c
                          c))))))
                    )
        )
; This condition specifies either Case 1 (crank above the
horizontal)
; or Case 2 (crank below the horizontal)
        (cond ((or (<= theta2 pi)(>= theta2 0.0))
            (setq theta3 (- psi beta)
               theta4 (- (* 2 pi) lambd beta)
            )
            )
            (t (setq theta3 (+ psi beta)
                  theta4 (-(+(* 2 pi) beta) lambd)
               )
               )
        )
        (setq pt3 (polar pt4 (- theta4 pi) 14))
; The bearings on the coupler are also drawn with a radius
equal to a
; twentieth of the length of crank.
     (command "circle" pt2 (/ 12 20))
     (command "circle" pt3 (/ 12 20))
; The polyline is defined with a varying thickness of the
crank from the small
; end bearing to the big end bearing depending on the length
of the crank.
     (command "pline" pt1 "w" (/ 12 200) (/ 12 50) pt2 pt3
     pt4 "")
; Since, only the polyline and bearings on the coupler are
to be deleted and
; redisplayed at a new position, 'LAST' option of the
'ERASE' command is used.
     (command "erase" "l" "")
     (command "erase" "l" "")
     (command "erase" "l" "")
; To increase the speed of the mechanism, the value that
increments the angle
; of the crank (theta2) is increased e.g. from 0.1 to 1.
        (setq theta2 (+ 0.5 theta2))
     )
)
```

APPENDIX A2.2

```
; TIPPER.LSP
; EdenLISP Program for the Tipper Mechanism.

; Declaration

llreal : tippernodes tipper
lreal  : pt1 pt2 pt3 pt4 pt5 pt6 pt7 pt8 pt9 size origin
real   : leverl bell boxb boxh stroke y x c1 c2 a1 b1 a2 b2
d e f a3 b3 basel
real   : a7 b7 boxang baser motion beta a4 b4

; Definitions and Expressions

tippernodes  = [pt1, pt2, pt3, pt7, pt6, pt9, pt8, pt3, pt2,
pt4, pt5]
tipper      = cedge(tippernodes)  ; topology

leverl    = 280.0
bell      = 50.0
boxb      = 440.0
boxh      = 100.0
stroke  = sqrt( a2*a2 + (b2-bell)*(b2-bell))
a1        = 0.0
b1        = 0.0
beta      = 23.0

y  = (c2-c1+2*a2*x)/(2*b2)
x  = - (e + sqrt (e*e - 4*d*f)) / (2*d)
d  = 4*a2*a2 + 4*b2*b2
e  = 4*a2*c1 - 4*a2*c2
f  = c2*c2 + c1*c1 + 4*b2*b2*c1 - 2*c1*c2
c1 = a1*a1 + b1*b1 - stroke*stroke
c2 = a2*a2 + b2*b2 - bell*bell

pt1  = [-440.0, 0.0]
pt2  = polar (pt1, dtor(beta), leverl)
pt3  = polar (pt4, ( atan((b2-b4)/(a2-a4)) + ( atan(bell /
bell))), sqrt(2*bell*bell))
pt4  = if stroke >= 240.0 then polar(pt1, dtor(beta),
(leverl - bell)) else [x, y]
pt5  = [0.0, 0.0]
pt6  = polar (pt3, ( boxang + dtor(90.0) + ( atan(basel/
boxh))), sqrt((boxh*boxh)+(basel*basel)))
pt7  = [-440.0, 109.405]
pt8  = polar (pt3, boxang, baser)
pt9  = polar (pt3, ( atan(boxh/baser) + boxang), sqrt((boxh*
boxh)+(baser*baser)))

a2      = projn(1, pt2)
b2      = projn(2, pt2)
a3      = projn(1, pt3)
```

```
b3     = projn(2, pt3)
a4     = projn(1, pt4)
b4     = projn(2, pt4)
basel  = (440.0 + a3)/(cos (boxang))
baser  = 440 - basel
a7     = projn(1, pt7)
b7     = projn(2, pt7)
boxang = atan ((b3-b7)/(a3-a7))

frame : tipping tipp

size    = [1, 1]
origin  = [-440.0, 0.0]

; Construct

tipp    = object(tipper, origin, size)

; Geometry

tipping = Wireframe(tipper, "pline")

;  Actions

motion  = ploop ("stroke", 191.7, 5.0, 245.0)
beta    = 25.0
beta    = 30.0
beta    = 35.0
beta    = 40.0
```

References

Ahola, T., Laitinen, E., Kujala J., & Wikstrom, K. (2008). Purchasing strategies and value creation in industrial turnkey projects. *International Journal of Project Management* 26: 87–94.

Al-Rousan, R., Sunar, M.S., & Kolivand, H. (2018). Geometry-based shading for shape depiction enhancement. *Multimedia Tools and Applications* 77: 5737–5766.

Alwan, Z.S., & Younis, M.F. (2017). Detection and prevention of SQL injection attack: A survey. *International Journal of Computer Science and Mobile Computing (IJCSMC)* 6(8): 5–17.

Autodesk. (2021). AutoCAD support and learning. Available: https://www.autodesk.com/products/autocad/overview?term=1-YEAR. Accessed: 23 June 2021.

Barari, A. (2009). Misinterpretation of geometric dimensioning and tolerancing in coordinate metrology. In: *CAT2009 Conference*.

Basoglu, N., Daim, T., & Sofuoglu, E. (2009. A decision methodology for customizing software products. *International Journal of Industrial and Systems Engineering* 4(5): 554–576.

Boole, G. (2011). *The Mathematical Analysis of Logic*. Project Gutenberg, North Carolina, CA.

Buchele, S.F., & Crawford, R.H. (2004). Three-dimensional halfspace constructive solid geometry tree construction from implicit boundary representations. *Computer-Aided Design* 36(11): 1063–1073. 10.1016/j.cad.2004.01.006.

Candi, M., & Beltagui, A. (2018). Effective use of 3D printing in the innovation process. *Technovation* 80: 63–73. 10.1016/j.technovation.2018.05.002

Carlota, V. (2019). Top 10 best CAD software for all levels. *3Dnatives*. Available: https://www.3dnatives.com/en/top10-cad-software-180320194/, Accessed: 23 June 2021.

Cartwright, A.J. (1994). Application of Definitive Scripts to Computer Aided Conceptual Design. PhD Thesis: University of Warwick, England.

Chen, Q., & Wang, G. (2003). A class of Bézier-like curves. *Computer Aided Geometric Design* 20: 29–39.

Cummins, K. (2010). The rise of additive manufacturing. *The Engineer*. Available: https://www.theengineer.co.uk/the-rise-of-additive-manufacturing/. Accessed: 22 June 2021.

Dano, H. (2012). Assessment of the awareness of structural computer aided design programs of Universities in Ghana. *European Journal of Social Sciences* 30(1): 41–47.

Du, T., Inala, J.P., Pu, Y., Spielberg, A., Schulz, A., Rus, D., Solar-Lezama, A., & Matusik, W. (2018). InverseCSG: Automatic conversion of 3D models to CSG trees. *ACM Transactions on Graphics*. 10.1145/3272127.3275006

Farin, G. (2002). *Curves and Surfaces for CAGD: A Practical Guide*. Morgan-Kaufmann, Burlington, MA. ISBN: 1-55860-737-4.

Fayolle, P.A., & Pasko, A.A. (2016). An evolutionary approach to the extraction of object construction trees from 3D point clouds. *Computer-Aided Design* 74: 1–17. 10.1016/j.cad.2016.01.001

Follmer, S.W., Leithinger, D., Olwal, A., Hogge, A., & Ishii, H. (2013). inFORM: Dynamic Physical Affordances and Constraints through Shape and Object Actuation. In: *UIST 2013 – Proceedings of the 26th Annual ACM Symposium on User Interface Software and Technology* (pp. 417–426). 10.1145/2501988.2502032.

Gherardini, F., Santachiara, M., & Leali, F. (2018). 3D virtual reconstruction and augmented reality visualization of damaged stone sculptures. *Materials Science and Engineering* 364: 012018. 10.1088/1757-899X/364/1/012018

Han, X., Ma, Y., & Huang, X. (2008). A novel generalization of Bézier curve and surface. *Journal of Computational and Applied Mathematics* 217: 180–193.

Hirz, M., Harrich, A., & Rossbacher, P. (2011). Advanced computer aided design methods for integrated virtual product development processes. *Computer-Aided Design and Applications* 8(6): 901–913. 10.3722/cadaps.2011.901-913.

Hsu, W.T., Lu, Y., & Ng, T. (2014). Does competition lead to customization? *Journal of Economic Behavior & Organization* 106: 10–28.

Kennedy, L. (2014). A brief history of AutoCAD. *Scan2D*. Available: https://www.scan2cad.com/blog/tips/autocad-brief-history/. Accessed: 19 April 2022.

Kiptiah binti Ariffin, M., Hadi, S., & Phon-Amnuaisuk, S. (2017). Evolving 3D Models Using Interactive Genetic Algorithms and L-Systems. In: Phon-Amnuaisuk, S., Ang, S.P., & Lee, S.Y. (eds.), *Multi-disciplinary Trends in Artificial Intelligence. MIWAI 2017. Lecture Notes in Computer Science* (vol. 10607, pp. 485–493). Springer, Cham. 10.1007/978-3-319-69456-6_40

Krysl, P., & Ortiz, M. (2001). Extraction of boundary representation from surface triangulations. *International Journal for Numerical Methods in Engineering* 50(7): 1737–1758.

Kushwaha, A., Verma, S.K., & Sharma, C. (2012). Analysis of the concerns associated with the rapid release cycle. *International Journal of Computer Applications* 52(12): 20–25.

Kutzner, S.C., & Kiesmüller, G.P. (2013). Optimal control of an inventory-production system with state-dependent random yield. *European Journal of Operational Research* 27: 444–452.

Lam, H.K.S., Ding, L., Cheng, T.C.E., & Zhou, H. (2019). The impact of 3D printing implementation on stock returns: A contingent dynamic capabilities perspective. *International Journal of Operations & Production Management* 39(6/7/8): 935–961.

Lee, S.H., Jeong, Y.S., & Kim, B.G. (2005). Sharing of steel bridge information using CAD system with ACIS solid modeler. *Journal of the Korean Society of Civil Engineers* 25(4A): 677–687.

Lin, J., & Hu, H. (2012). Software development scheme based on AutoCAD for graphical model design of smart substation. *Electric Power Automation Equipment* 32(9): 142–148.

Liu, C.Y.D., & Ridgway, K. (1995). A computer-aided inventory management system – Part 1: Forecasting. *Journal Integrated Manufacturing Systems* 6(1): 12–21.

Liu, X., Wang, W., & Peng, R. (2015). An integrated production, inventory and preventive maintenance model for a multi-product production system. *Journal Reliability Engineering and System Safety* 137: 76–86.

Madsen, D.A., & Madsen, D.A. (2012). *Engineering Drawing and Design*. Cengage Learning, Boston, MA. ISBN: 1111309574.

Mahmudova, S.J. (2018). Development tendencies for programming languages. *Journal of Engineering Technology* 9(1). Available: https://www.researchgate.net/publication/326111597, Accessed: 24 May 2022.

Maqsood, S., Abbas, M., Miura, K.T., Majeed, A., & Iqbal, A. (2020). Geometric modeling and applications of generalized blended trigonometric Bézier curves with shape parameters. *Advances in Differential Equations* 2020(1): 1–18.

Maurer, W.D. (1972). *The Programmers Introduction to LISP*. The Book Service Ltd, Colchester, UK. ISBN-13: 978-0356039800.

Mell, M., & Monroy, F. (2018). A gradient-based, GPU-accelerated, high-precision contour-segmentation algorithm with application to cell membrane fluctuation spectroscopy. *PLoS One* 13(12): e0207376. 10.1371/journal.pone.020737.

Misra, S.C., & Singh, V. (2015). Conceptualizing open agile software development life cycle (OASDLC) model. *International Journal of Quality & Reliability Management* 32(3): 214–235.

Narayan, K.L., Rao, K.M., & Sarcar, M.M.M. (2008). *Computer Aided Design and Manufacturing* (p. 3). New Delhi: Prentice Hall of India. ISBN: 978-8120333420.

Norman, E., Riley, J., Urry, S., & Whittaker, M. (1990). *Advanced Design Technology*. Harlow, UK: Longman.

Nyemba, W.R. (1999). Introducing AutoLISP in mechanism design and analysis. *Proceedings of the Zimbabwe Institution of Engineers* 1(2): 29–34.

Nyemba, W.R. (2000). Three-dimensional modelling of a connecting rod using constructive solid geometry. *African Journal of Science and Technology* 1(1): 83–90.

Nyemba, W.R. (2013). Modelling and simulation of a dump truck tipping mechanism using EdenLISP. *Journal of Science, Engineering and Technology* 1(2): 1–6.

Nyemba, W.R., & Lambu, J. (2006). A CAD system for the automatic generation of bills of materials for a motor vehicle manufacturing company. *Proceedings of the Zimbabwe Institution of Engineers* 4(1): 32–39.

Nyemba, W.R. & Mbohwa, C. (2016). Customization of the CAD software in a typical drawing office for a power and electricity distribution company in Zimbabwe. In: *2016 IEEE International Conference on Industrial Engineering and Engineering Management*, 4–7 December 2016 (pp. 696–700), Bali, Indonesia. ISBN: 978-1-5090-3665-3. IEEE.

Nyemba, W.R., & Mbohwa, C. (2017). Design of a 10-digit inventory codification system for a tube and pipe manufacturing company in Zimbabwe. *Procedia Manufacturing* 8: 503–510.

Nyemba, W.R., Mbohwa, C., & Carter, K.F. (2021). *Bridging the Academia Industry Divide: Innovation and Industrialisation Perspective using Systems Thinking Research in Sub-Saharan Africa. EAI/Springer Innovations in Communication and Computing Series*. Springer, Cham, Switzerland. ISBN: 978-3-030-70492-6; ISBN: 978-3-030-70493-3 (eBook).

Panneerselvam, R. (2012). *Production and Operations Management*, 3rd edition. Prentice Hall, New Jersey, USA. ISBN: 978-81-203-4555-3.

Pottmann, H., Brell-Cokcan S., & Wallner, J. (2007). Discrete Surfaces for Architectural Design. Archived 2009-08-12 at the Wayback Machine. In *Curve and Surface Design*, Chenin, P., Lyche, T., & Schumaker, L.L. (eds.), Nashboro Press, pp. 213–234. ISBN: 978-0-9728482-7-5.

Quintana, V., Rivest, L., Pellerin, R., & Kheddouci, F. (2012). Re-engineering the engineering change management process for a drawing-less environment. *Computers in Industry* 63: 79–90.

Remondino, F., & El-Hakim, S. (2006). Image-based 3D modelling: A review. *The Photogrammetric Record* 21(115): 269–291.

Sharma, A., & Singh, B.J. (2020). Evolution of industrial revolutions: A review. *International Journal of Innovative Technology and Exploring Engineering* 9(11): 66–73.

Simmons, C.H., Maguire, D.E., & Phelps, N. (2020), Principles of First and Third Angle Orthographic Projection Chapter 6. In: Simmons, C.H., Maguire, D.E., & Phelps, N. (eds.), *Manual of Engineering Drawing*, 5th edition (pp. 49–70). Butterworth-Heinemann.

Sivasankaran, P. & Radjaram, B. (2020). Recent applications of 3D printing and its challenges – A review. *International Journal of Engineering and Technical Research* V9(06): 396–399. 10.17577/IJERTV9IS060240.

Stites, W.M., & Drake, P. (1999). Geometric Dimensioning and Tolerancing. In: *Dimensioning and Tolerancing Handbook*, Drake, P. (ed.), ISBN-13: 9780070181311.

Stroud, I. (2006). *Boundary Representation Modelling Techniques*. Springer-Verlag, London. ISBN-10: 0-387 84628-312-4.

Touretzky, D.S. (1990). *COMMON LISP: A Gentle Introduction to Symbolic Computation*. The Benjamin/Cummings Publishing Company, Inc., California, USA. ISBN: 0-8053-0492-4.

van Horenbeek, A., Buré, J., Cattrysse, D., Pintelon, L., & Vansteenwegen, P. (2013). Joint maintenance and inventory optimization systems: A review. *International Journal of Production Economics* 143: 499–508.

van Rossen, S., & Baranowski, M. (2011). Real-Time Constructive Solid Geometry. In: Ansari, M. (ed.), *Game Development Tools* (pp. 79–96). CRC Press. ISBN: 9781439867723.

Vasiou, E., Shkurko, K., Mallett, I., Brunvand, E., & Yuksel, C. (2018). A detailed study of ray tracing performance: render time and energy cost. *The Visual Computer* 34: 875–885.

Verma, V., & Walia, E. (2010). 3D rendering – Techniques and challenges. *International Journal of Engineering and Technology* 2(2): 72–77.

Vinodh, S., Devadasan, S.R., & Shankar, C. (2010). Design agility through computer aided design. *Journal of Engineering, Design and Technology* 8(1): 94–106.

Vinogradov, O. (2020). *Fundamentals of Kinematics and Dynamics of Machines and Mechanisms*. CRC Press, Boca Raton, FL. ISBN: 9780367398323.

Wang, T., Liang, C., An, Y., Xiao, S., Xu, H., Zheng, M., Liu, L., Wang, G., & Nie, L. (2020). Engineering the translational machinery for biotechnology applications. *Molecular Biotechnology* 62(4): 219–227.

Wang, Y., Kandampully, J., & Jia, H.M. (2013). Tailoring customization services. *Journal of Service Management* 24(1): 82–104.

Zhuo, Y., Wu, Y.J., & Peng J. (2012). Design and simulation of 3D layout for MID based on open CASCADE. *Advanced Materials Research* 479–481: 1978–1981

Zuo, B.Q., Huang, Z.D., Wang, Y.W., & Wu, Z.J. (2014). Isogeometric analysis for CSG models. *Computer Methods in Applied Mechanics and Engineering* 285: 102–124. 10.1016/j.cma.2014.10.046.

Index

Pages in *italics* refer figures; pages in **bold** refer tables.